*Series Editor*
*Piotr Breitkopf*

# Topology Optimization in Engineering Structure Design

Weihong Zhang
Jihong Zhu
Tong Gao

ISTE PRESS

ELSEVIER

First published 2016 in Great Britain and the United States by ISTE Press Ltd and Elsevier Ltd



ISTE Press Ltd
27-37 St George's Road
London SW19 4EU
UK

www.iste.co.uk

Elsevier Ltd
The Boulevard, Langford Lane
Kidlington, Oxford, OX5 1GB
UK

www.elsevier.com



British Library Cataloguing-in-Publication Data
A CIP record for this book is available from the British Library
Library of Congress Cataloging in Publication Data
A catalog record for this book is available from the Library of Congress
ISBN 978-1-78548-224-3

# Contents

# Introduction

## I.1. Overview and motivation

> After the Second World War, the rapid developments of the aircraft and aerospace industries brought great engineering challenges and academic interests in lightweight and high performance structure design. Structural optimization techniques are becoming increasingly important to satisfy complicated engineering requirements.
>
> Strive to save the weight by each gram.
>
> – AVIC, China
>
> For aircraft or aerospace vehicles, weight is a critical determinant of performance, payload capacity, maneuverability and range.
>
> – US National Research Council

In recent decades, structural optimization methods have gained great progress with the increasing performances of computers and computing algorithms. Solutions of practical and complicated optimization problems undergoing complex loading conditions are made possible to satisfy severe multidisciplinary design performances. Among others, topology optimization has become one of the most promising techniques.

In the 21st Century, many new aircraft and aerospace projects are being set up in China. This brings great challenges in developing innovative design methodology and dealing with new scientific and technical problems issued from the complicated engineering practices. Nowadays, industrial applications are becoming one of the most important challenges in the engineering design community of structure optimization, especially topology optimization for aircraft or aerospace structure systems. Within this scope, fundamental and innovative research works are carried out. The research team of Engineering Simulation & Aerospace Computing in Northwestern Polytechnical University in China is becoming one of the most active research groups in this discipline and is motivated to develop the following techniques and solutions.

In 2003, we started to optimize the wing structure of an aircraft to satisfy the specific static and dynamic performances. The global model consists of less than 5,000 shell, rib and beam elements. Later, the work was focused on the optimization of the composite and honeycomb structures of a large airborne radome. As mechanical performances and electromagnetic functionality were contradictory in design, both of them were optimized simultaneously based on multiobjective optimization methods.

Since 2005, studies have been focused on the shape optimization of aero-engine structures, for example turbine disks, blades, shafts, elastic supports and nozzle parts that belong to a kind of complicated curved structure working under aerodynamic loads, rotating inertial loads and high temperature conditions. Typically, shape optimization of the cutouts on the thin-walled curved panels was a challenging issue because it could not be directly dealt with using traditional shape optimization methods. It is required that the cutout boundary should be kept on the curved surface whatever the design modification. Thus, both a new mapping method and Boolean operation method were developed to define design variables of the cutout in the intrinsic coordinate system of the curved surface. Mechanical and thermal stresses were considered in this work. Shape optimization resulted in a stress-level reduction of about 30% and a weight saving of more than 10%.

Almost at the same time, we were motivated by the conceptual design of new aircraft structures. Based on fruitful discussions with aircraft design engineers, topology optimization of large-scale aircraft structures was carried out with nearly one million design variables and dozens of complex loading conditions. In particular, design-dependent loads and mass constraint with multiple materials should be treated properly in topology optimization to achieve clear structural configurations.

At the beginning of 2006, we were motivated by the simultaneous optimization of payloads and supporting structures after a visit by some aerospace engineers. The aim was to integrate geometrical packing optimization with topology optimization for the system of large launch vehicles. An integrated design methodology of a multicomponent system was thus developed. The supporting structure configuration and layout of satellites are optimized to improve the global dynamic performance. Shape and sizing optimizations were further used to detail the structural design for the strength and buckling requirements. In consequence, the dynamic performance is increased by 17% with a weight saving of more than 7% compared to the initial design.

Based on the above practices, we have realized that structure topology optimization has become essentially important to promote the frontier industrial developments, especially for the design of advanced aircraft and aerospace systems. The purpose of this book is to present recent achievements of topology optimization. Methods, techniques and applications summarized in the book will hopefully be destined to provide rich illustrations for researchers and engineers working in the field of structure design.

## I.2. Basic engineering optimization methodologies

Structure optimization methods are basically classified into three categories: sizing optimization, shape optimization and topology optimization. Sizing optimization is a classical method and easy to conduct by choosing cross-sectional dimensions of trusses, beams and frames, or the thicknesses of membranes, plates and shells as design variables, as shown in Figure I.1. Sizing optimization can be regarded as a detailed design procedure of the structural model involving a large number of design variables. It has been developed maturely and is becoming the most popular method in engineering community.

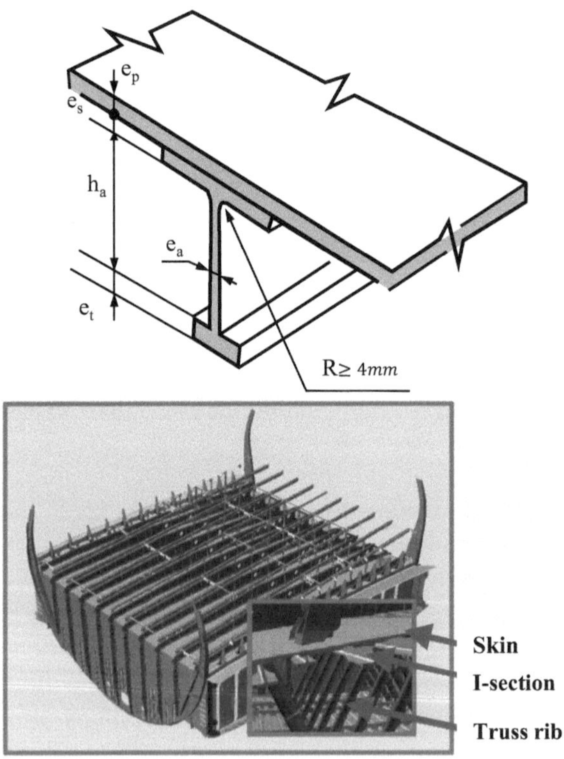


**Figure I.1.** *Sizing optimization for aircraft structure members [RAD 02]*

Shape optimization aims at designing structural boundaries or holes in a structure, as shown in Figure I.2. This method can be used practically to ameliorate local performances such as stress distribution. There often exist a small number of geometric design variables owing to boundary parameterization. As the boundary perturbation directly changes the geometrical model, shape sensitivity analysis with respect to geometric design variables is always a problem to be carefully considered. Obviously, both sizing and shape optimization methods are detailed design procedures without changing the specific topology of a structure.

**Figure I.2.** *Shape optimization of cutouts in a thin-walled structure*

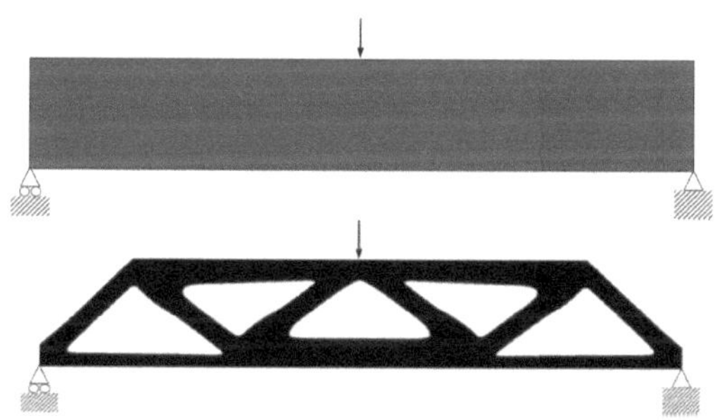

**Figure I.3.** *A typical topology optimization problem of MBB-beam [BEN 03]*

Topology optimization aims at finding an optimal solid-void pattern of the material layout over a specific design domain with given boundary conditions, as shown in Figure I.3. This method is often used at the conceptual design stage to optimize global performances such as the rigidity and natural frequencies of a structure. Topology optimization was originally a 0-1 discrete problem. The major challenge is the solution of a large-scale integer programming problem. The high computing cost of this kind of problem typically precludes the use of gradient-free algorithms. The successful application of Lagrangian duality to the large-scale integer problem was found in the work of Beckers [BEC 97, BEC 99] who proposed a dual method to deal with discrete design variables. Most approaches have been proposed to deal with the problem as a continuous one since the pioneering work by Cheng and Olhoff [CHE 81] and Bendsøe and Kikuchi [BEN 88]. To have a comprehensive understanding of the state of the art, one can refer to literature surveys and books by Bendsøe [BEN 95, BEN 02], Eschenauer and Olhoff [ESC 01] and Bendsøe and Sigmund [BEN 03].

Several representative methods have been proposed so far. For example, the homogenization-based method [BEN 88, GUE 90, SUZ 91, ALL 04a] describes the structural material layout with microstructures, as shown in Figure I.4. Meanwhile, the equivalent material properties of each microstructure, for example elastic moduli, are calculated using the homogenization method. Topology optimization is processed by modifying the dimension parameters of each microstructure iteratively. However, the mathematical complexity of the homogenization process prevents the general application of this method.

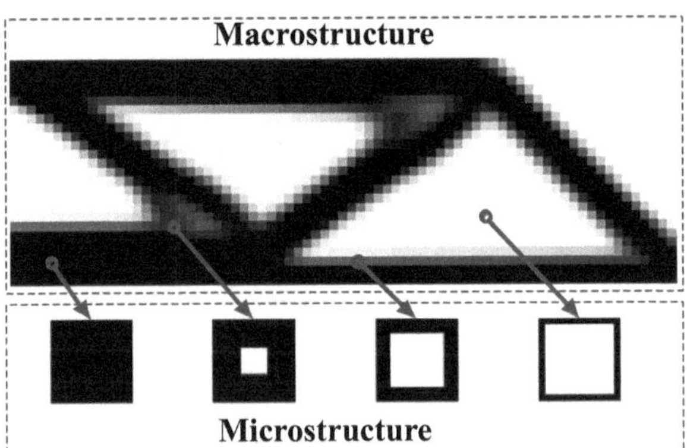


**Figure I.4.** *Material layout described with the homogenization-based method*

Among others, solid isotropic material with penalty (SIMP) is the most popular method in topology optimization [BEN 89, BEN 99, ZHO 91, ROZ 01a]. It proceeds by penalizing exponentially isotropic material in terms of element pseudo-density variables defined by a power-law

$$E_i = E_0 \eta_i^p \qquad [I.1]$$

where $E_i$ is the elastic modulus of the $i$th element. $E_0$ is the elastic modulus of the solid material. $\eta_i$ and $p$ are the so-called pseudo-density and penalty factor, respectively. Compared with the amount of material for each element, a very low stiffness will be obtained even when element pseudo-density variables take intermediate values between 0 and 1. The effect of the penalty factor will push the pseudo-density toward 0 and 1 during the optimization. In the work of Bendsøe and Sigmund [BEN 99], the power-law approach was proved theoretically provided that the penalty factor satisfies the Hashin–Shtrikman bounds.

To obtain a purely or nearly 0-1 material layout, strong penalty with a great value of $p$ is suggested. Unfortunately, as the derivative of $E_i$ with respect to $\eta_i$ is zero at $\eta_i = 0$, the presence of the so-called gray elements may occur in a numerical solution. To ensure the numerical stability, rational approximation of material properties (RAMP) was proposed by Stolpe and Svanberg [STO 01] as an alternative model.

$$E_i = \frac{\eta_i}{1+q(1-\eta_i)} E_0 \qquad [I.2]$$

where $q$ is the penalty factor.

In comparison, SIMP and RAMP are similar to each other. However, RAMP is sensitive to $\eta_i$=0 in the sense that $E_i$ has a non-zero derivative at $\eta_i = 0$.

The evolutionary method is an engineering approach. It is based on the intuitive concept that inefficient materials are gradually removed from the design domain to approach the optimal topology. Among others, evolutionary structural optimization (ESO) developed by Xie and Steven [XIE 96], Li *et al.* [LI 99] and Kim *et al.* [KIM 03] is a typical evolutionary approach. In most cases, optimal topologies are generated by deleting the set of elements with low strain energy values from the entire design domain systematically. The element efficiency evaluated from sensitivity analysis is used as an index of element deletion. As the ESO is devised as a unidirectional scheme only for removing elements, the restitution of the removed elements will be, however, unallowable during the iteration. Later, an improved bidirectional procedure named bidirectional evolutionary structural optimization (BESO) was proposed by Querin and Young [QUE 00] and Yang *et al.* [YAN 99a, YAN 99b]. Materials are allowed to be added in those void areas with the highest efficiency, but it is required that an initial design configuration connecting the boundary conditions and loading locations should be specified *a priori.*

Both ESO and BESO have the advantage of conceptual simplicity. Moreover, Tanskanen [TAN 02] proved that, in some particular situations, these approaches basically correspond to a sequential linear programming approximate method. However, Sigmund [SIG 01] indicated that it is questionable to extend these approaches to other design cases such as multiconstraints and multiphysics problems. A critical view given by Zhou and Rozvany [ZHO 01], Rozvany [ROZ 01b] also indicated the existence of some numerical failures. In particular, neither the stress level nor the sensitivity values used till now has been able to describe exactly the criterion of the element deletion/growth when the latter causes a significant variation of the objective function [ZHU 07]. Nevertheless, the ESO method was also defended against the criticism [EDW 07, HUA 08, TAN 02, ROZ 02a, ROZ 02b, ROZ 04].

Other topology optimization methods were also proposed. For example, the bubble method developed by Eschenauer *et al.* [ESC 94] introduced new holes (or bubbles) into the design domain. The contour and position of the holes were designed in the way of shape optimization. The topology description function and level set method

[WAN 03, ALL 04b, DER 04, MEI 04a, MEI 04b] were developed to describe the structure as a high dimension level-set function.

It should be mentioned that alternating solid and void elements over the design domain often occurs in topology optimization. This phenomenon behaves in a checkerboard fashion and is mesh dependent. According to Jog and Haber [JOG 96], it was due to the finite element approximation or design optimization criteria. From this viewpoint, Rodrigues and Fernandes [ROD 95] improved the interpolation accuracy by means of high-order elements in thermo-elastic optimization problems. However, the computing cost increases dramatically together with the number of degrees of freedom of the structural system. Later, Sigmund and Petersson [SIG 98] developed the filtering scheme to smooth the sensitivities of the objective functions over the considered element and its eight neighbors based on image filtering techniques. However, this sensitivity filter is not appropriate for the searching strategies because the modified sensitivities do not completely correspond to the objective function and may lead to some divergence problems. As a result, further developments are being made on the density filter by Bruns and Tortorelli [BRU 01] and Bourdin [BOU 01]. The modifications are directly implemented on the updated design variables. More descriptions and improvements on density filter can be found in the works of Wang and Wang [WAN 05], Sigmund [SIG 06], Sigmund [SIG 07] and Lemaire *et al.* [LEM 07].

Alternatively, Haber *et al.* [HAB 96] proposed the perimeter control method to control the checkerboard pattern and detailed structures between solid and void elements. Zhang and Duysinx [ZHA 03] also proposed an improved perimeter control of quadratic form in consistence with the dual approach. Checkerboard control in the framework of ESO/BESO was discussed by Yang *et al.* [YAN 02].

The idea of topology optimization has been extended to different territories. Numerical results show that a variety of problems including maximization designs of structural stiffness [SIG 01a], natural frequency [PED 00], buckling loads [ZHO 04], heat conduction [GER 06], CFD channel flow [GER 05] etc., can be solved. Furthermore, the concept of topology design domain is extended by introducing structural supports and joints modeled with spring elements. Typical results presented by Jiang and Chirehdast [JIA 97], Buhl [BUH 01], Zhu and Zhang [ZHU 06a] mainly covered problems of structural stiffness, natural frequency as well as compliant mechanism. Other extended patterns of topology optimization were developed to design the microstructures [SIG 99, ZHA 06] and to deal with design-dependent load problems [CHE 01, BRU 05, GAO 08].

When the eigenvalue problems like natural frequencies and buckling loads of a structure are maximized with the SIMP model, an important issue concerns the artificial modes or localized deformations. This issue takes place in low-density areas where elements take the minimum pseudo-density values. Compared with the solid region, these areas are too compliant to support themselves. Neves *et al.* [NEV 95] investigated this phenomenon when optimizing the structural buckling loads. Pedersen [PED 00] and Bruyneel and Duysinx [BRU 05] improved SIMP interpolation model after analyzing the artificial modes numerically in natural frequency maximization and self-weight loading problems, respectively. By analyzing the material properties of the elements in low-density areas, Zhu *et al.* [ZHU 07] used the equivalent material properties of the orthotropic cellular microstructures that could be effective in avoiding the artificial modes.

Topology optimization has stepped into its rapid developing age. Excellent works gain great success in both theoretical studies and practical applications. Recent literature surveys are given by Guo and Cheng [GUO 10], Deaton and Grandhi [DEA 14], Sigmund and Maute [SIG 13] Zhang *et al.* [ZHA 11] and Zhu *et al.* [ZHU 16].

## I.3. Layout of the book

This book consists of seven chapters mainly summarizing the work of the authors' research team. A brief introduction of the background and motivation is presented first. Then the state of art of the related techniques and their applications is reviewed. In the following chapters, contents are related to standard material layout design with topology optimization, low-density areas in topology optimization, dynamic and thermal-elastic design of topology optimization, integrated layout design of multicomponent system and topology optimization with constraints on multifastener joint loads. Finally, potential applications of topology optimization, such as shape preserving design, smart structure design, structural feature design and additive manufacturing, etc., are also addressed to provide a forward-looking perspective.

# 1

# Standard Material Layout Design

## 1.1. Basic formulations of topology optimization

In most engineering applications, topology optimization has been recognized as an effective approach for conceptual design. Topology optimization results were considered as a design of the most effective load carrying path, while the structural details in the design domain, such as structural chamfers and fillets, stiffeners, joints and cross-sections, were designed in the following shape and sizing optimization procedure.

A basic topology design procedure can be illustrated with the help of a typical application of a large cargo aircraft pylon design. The pylon shown in Figure 1.1 hangs the turbine engine or other payloads to the aircraft wing. The structure will be optimized to produce a stiff and lightweight design. First, the pylon is assigned as the design domain that is discretized by refined solid finite elements. Two hanging positions, i.e. the tip and lower lugs, are considered as non-designable components. The design domain and its finite element model (240,000 elements) are shown in Figure 1.2. The thrust and weight of the turbine engine are considered as applied loads on both hanging positions. The total structure is clamped onto the wing at two fixation positions.

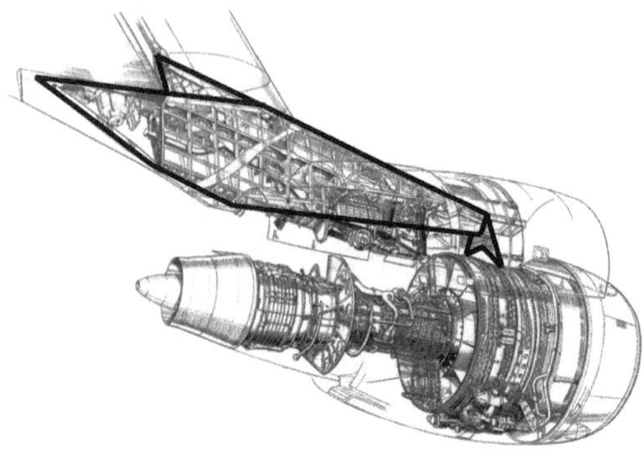

**Figure 1.1.** *An aircraft pylon (http://www.flightglobal.com) hanging a turbine engine*

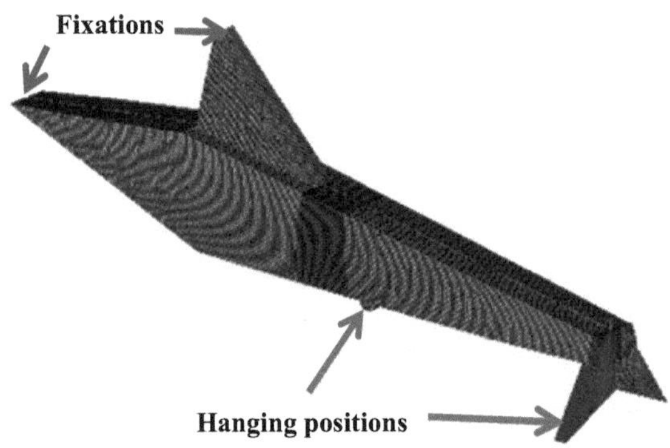


**Figure 1.2.** *The design domain and its finite element model*

Topology optimization of this problem can be formulated in the following way. First, design variables are defined by pseudo-densities describing the material distributions.

$$\text{find:} \quad \boldsymbol{\eta} = \{\eta_i\} \quad 0 < \eta_i \leq 1, \; i = 1, 2, \ldots, n_e \tag{1.1}$$

where $\eta_i$ refers to the pseudo-density variable of element $i$ representing a solid or void when it takes the value of 1 and 0, respectively. $n_e$ is the number of pseudo-density variables. Practically, to avoid singularities of computed element properties, for example the element stiffness matrix, a small non-zero value is often assigned as the lower bound of $\eta_i$.

Suppose the design objective is to maximize the global structural stiffness, which is normally evaluated as the minimization of the mean compliance.

$$\min: \quad C = \frac{1}{2}\mathbf{F}^{\mathrm{T}}\mathbf{U} = \frac{1}{2}\left(\mathbf{F}^{\mathrm{m}} + \mathbf{F}^{\mathrm{g}}\right)^{\mathrm{T}}\mathbf{U} \qquad [1.2]$$

where $C$ is the strain energy in terms of the external load vector $\mathbf{F}^{\mathrm{m}}$, the self-weight load vector $\mathbf{F}^{\mathrm{g}}$ and the nodal displacement vector $\mathbf{U}$. $\mathbf{F}$ is the total nodal load vector. It is important to note that $\mathbf{F}^{\mathrm{g}}$ is design dependent and follows the general rule of no material, no load. The finite element equilibrium equation corresponds to

$$\mathbf{F}^{\mathrm{m}} + \mathbf{F}^{\mathrm{g}} = \mathbf{K}\mathbf{U} \qquad [1.3]$$

Here, $\mathbf{K}$ is the structural global stiffness.

Moreover, a material volume constraint is needed in practical engineering design.

$$\text{s.t.} \quad V = \sum_i \eta_i V_{i0} \leq V_{\mathrm{U}} \qquad [1.4]$$

$V$ and $V_{\mathrm{U}}$ denote the total volume of the material and the upper bound, respectively. The combination of equation [1.2] with equation [1.4] constitutes the standard topology optimization formulation.

In this formulation, apart from SIMP and RAMP, a polynomial interpolation model [ZHU 09] is introduced to consider the design-dependent effect of self-weight. In this context, different penalties are used to match both variations of the self-weight and stiffness.

$$\begin{aligned} \rho_i &= \eta_i \rho_{i0} \\ E_i &= P(\eta_i) E_{i0} = \left((1-w)\eta_i^p + w\eta_i\right) E_{i0} \end{aligned} \qquad [1.5]$$

where $\rho_{i0}$ and $E_{i0}$ are the specific density and elastic modulus of solid material, respectively. $P(\eta_i)$ can be considered as an extended SIMP with $w$ and $p$ being constant parameters of the interpolation. Details of this interpolation model are discussed in the following chapters.

To solve the optimization problem, sensitivities of the objective function and constraint function with respect to the pseudo-density variables are needed.

$$\begin{aligned} \frac{\partial V}{\partial \eta_i} = V_{i0}\,;\ \frac{\partial C}{\partial \eta_i} &= \frac{1}{2}\left[\frac{\partial \left(\mathbf{F}^{\mathrm{g}}\right)^{\mathrm{T}}}{\partial \eta_i}\mathbf{U} + \left(\mathbf{F}^{\mathrm{m}} + \mathbf{F}^{\mathrm{g}}\right)^{\mathrm{T}} \frac{\partial \mathbf{U}}{\partial \eta_i}\right] \\ &= \frac{1}{2}\frac{\partial \left(\mathbf{F}^{\mathrm{g}}\right)^{\mathrm{T}}}{\partial \eta_i}\mathbf{U} + \frac{1}{2}\left(\mathbf{F}^{\mathrm{m}} + \mathbf{F}^{\mathrm{g}}\right)^{\mathrm{T}} \mathbf{K}^{-1}\left(\frac{\partial \mathbf{F}^{\mathrm{g}}}{\partial \eta_i} - \frac{\partial \mathbf{K}}{\partial \eta_i}\mathbf{U}\right) \end{aligned} \qquad [1.6]$$

If polynomial interpolation of equation [1.5] is used, we can easily obtain the above derivative.

To mitigate possible numerical instabilities, such as mesh-dependency, checkerboard patterns, the standard sensitivity filter proposed by Sigmund and Petersson [SIG 98] is the most popular method and used here.

$$\left(\frac{\partial C}{\partial \eta_i}\right)^* = \frac{\sum_j \left\{\eta_j \left[r_{\min} - \mathrm{dis}(j,i)\right] \frac{\partial C}{\partial \eta_j}\right\}}{\eta_i \sum_j \left[r_{\min} - \mathrm{dis}(j,i)\right]} \qquad [1.7]$$

$$\eta_j \in \left\{\eta_j \middle| \mathrm{dis}(j,i) \le r_{\min}\right\}$$

where dis($j$,$i$) refers to the distance between the $j$th and $i$th density points. All the density points located within the predefined distance $r_{\min}$ are accounted for in the filtering scheme.

Numerical experiences indicate that topology optimization problems can be solved with different algorithms, such as Conlin [FLE 86, FLE 89]), MMA (method of moving asymptotes, [SVA 87], SQP (sequential quadratic programming, see [NOC 99]), MDQA (method of diagonal quadratic approximations, [ZHA 97]) and GCMMA (global convergent version of MMA, [SVA 95, SVA 07]). For example, the GCMMA was implemented in BOSS-Quattro which is a general purpose optimization platform [RAD 02]. This algorithm is based on the approximation of a function expressed as

$$\mathbf{x}=\left[x_1,x_2,\ldots,x_n\right]^{\mathrm{T}}$$

$$g(\mathbf{x})\approx g\left(\mathbf{x}^{(k)}\right)+\sum_i p_i^{(k)}\left(\frac{1}{u_i^{(k)}-x_i}-\frac{1}{u_i^{(k)}-x_i^{(k)}}\right)+\sum_i q_i^{(k)}\left(\frac{1}{x_i-l_i^{(k)}}-\frac{1}{x_i^{(k)}-l_i^{(k)}}\right) \qquad [1.8]$$

Parameters $p_i^{(k)}$ and $q_i^{(k)}$ are computed based on the first-order derivatives of the function. Asymptotes $u_i^{(k)}$ and $l_i^{(k)}$ are two positive parameters that should be updated on the basis of the rule proposed by Svanberg [SVA 95] to ensure the global convergence of the algorithm.

The benefits of GCMMA are obvious. As it holds the property of non-monotonicity, convexity and separability in terms of variables, it is thus helpful in solving optimization problems involving non-monotonous functions.

For the optimization problem shown in Figure 1.2, the final design obtained using GCMMA is shown in Figure 1.3. The material distribution of the optimized design indicates the most effective load carrying path that can be used as an important reference for the further detailed design.

**Figure 1.3.** *Topology optimization design of the aircraft pylon*

## 1.2. Typical applications of standard topology optimization

In the work of Remouchamps *et al.* [REM 11], the conceptual design for the pylon of Airbus A350 is obtained using standard topology optimization implemented in the SAMCEF platform. The global design domain was meshed with tetrahedral elements. To ensure the symmetry, design variable linking was used. The design domain and optimized design are shown in Figure 1.4.

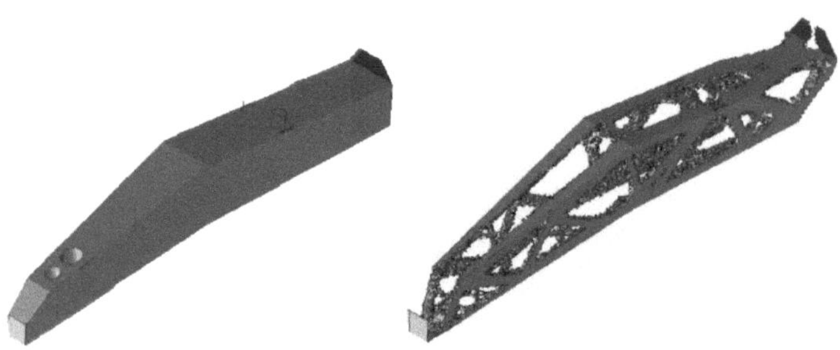

**Figure 1.4.** *Topology optimization design of the Airbus A350 pylon [REM 11]*

Another example is the rear fuselage design of a large cargo aircraft. Due to the structural symmetry, only half of the finite element model is shown in Figure 1.5. It is a relatively huge design with

460,000 design variables and 22 load cases including the inner pressure, outer fight loads, loads from the rear cabin door and the vertical tail. The whole model is fixed on the section connected to the middle fuselage. The topological design is shown in Figure 1.6 after more than 50 design iterations. Clear structural pattern including several reinforced frames, stringers, etc. can be found.

**Figure 1.5.** *Finite element model of the rear fuselage. For a color version of this figure, see www.iste.co.uk/zhang/topology.zip*

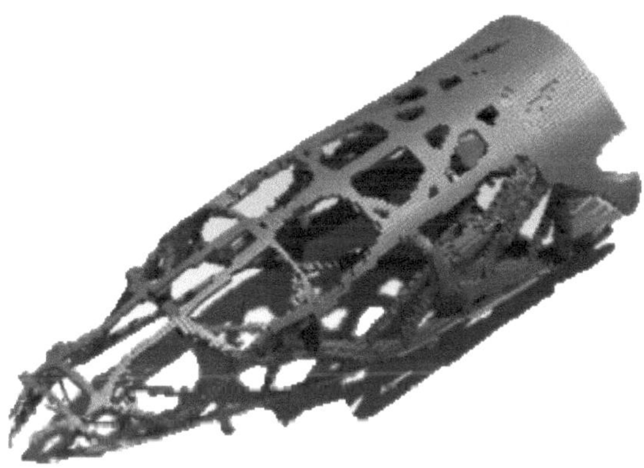

**Figure 1.6.** *Topology optimization design of the rear fuselage*

The optimized design obtained in the standard topology optimization is only a rough material distribution [MAU 04]. Further detailed designs are necessary for the purpose of engineering applications. In this case, the optimized design will be post-processed as engineering models accordingly by adding more engineering features considering manufacturing, assembling and functional purposes. This procedure can be done in some general purpose CAD platforms by skilled designers. The detailed configuration and performances of the final model will, however, depend upon the subjectivity and experience of the designer. Some post-processing schemes that automatically generate a smooth boundary of the topological design were also presented in the work of Bendsøe and Sigmund [BEN 03] and Sigmund [SIG 07], where the iso-density curve or surface on the boundary are identified and used. The performances of the smoothened model were proved to be very close to the topological design in many works. Moreover, it was also found that some nodal pseudo-density variable-based topology optimization methods favor the post-processing [GUE 04].

Once the optimized design has been post-processed into a detailed engineering model, subsequent shape and sizing optimizations are needed to improve the structural performances that are not fully considered in topology optimization, for example, local stress, buckling, dynamic response, etc.

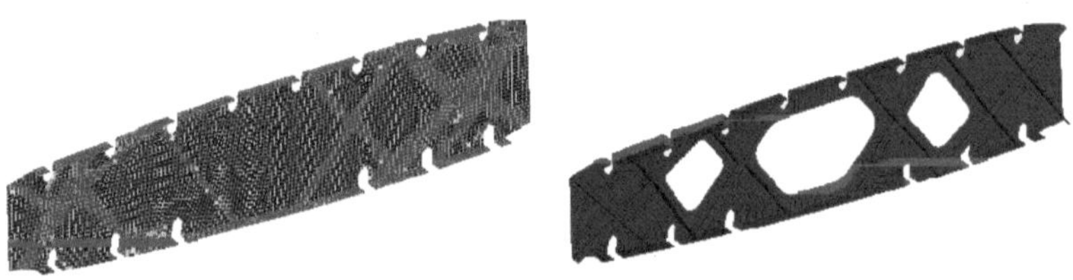

**Figure 1.7.** *Topologically optimized design and the rebuilt model [KRO 02]*

Typical design procedures can be found in the works presented by Krog *et al.* [KRO 02, KRO 04], where the least weight design of Airbus A380 wing structures was implemented. As shown in Figure 1.7, according to the topology design of a wing rib, the global structural configurations were rebuilt and optimized in detail. Some differences in structural configurations were found between the topologically optimized design and the rebuilt design. This was mostly due to the consideration of detailed requirements. However, in some other cases, this could also be a compromise to existing design traditions.

One of the most important applications of topology optimization was also presented by Krog *et al.* [KRO 02]. As shown in Figure 1.8, the leading edge ribs of Airbus A380 were optimized with significant effects of weight saving.

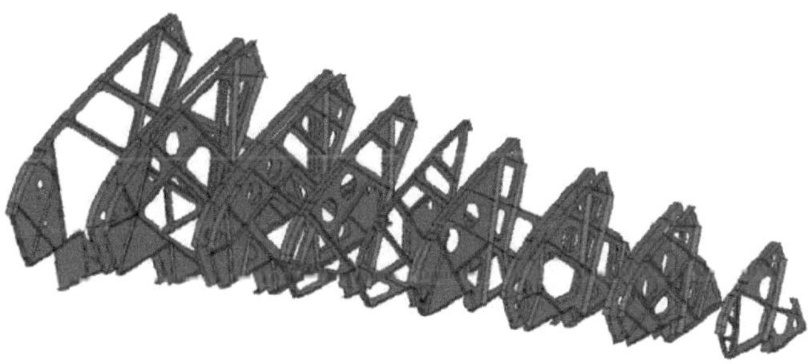

**Figure 1.8.** *Topology optimization of Airbus A380 leading edge ribs*

Attention should be paid to the finite element model and the design domain of topology optimization. Usually, practical topology optimization is to design some structural components which are only a fraction of the total aircraft model. In this case, the external loads are applied on the total model, and the design domain is then discretized with refined finite element mesh to maintain the quality of topology optimization. This is a kind of global model approach. However, a refined finite element model of a total aircraft will be on a massive scale, consisting of millions of elements. Topology optimization using global model analysis is computationally expensive. For this reason, substructuring techniques have been popularly used [KRO 04]. After the analysis of the global model, the target component and the interface loads are extracted from the global model. Topology optimization is then carried out with the local model and the interface loads. The substructure-based methods are obviously approximate schemes as the interface loads are not updated during the structural topology evolution. However, it was a practical and effective way in aircraft structure analysis and design.

## 1.3. Topology optimization of cellular materials and structures

### 1.3.1. *Homogenization method and material microstructure designs*

Cellular solids are ultra-lightweight materials widely applied in aerospace and automotive industries due to their particular multifunctional properties such as energy absorption, thermal isolation, anti-impact, etc. The structural efficiency can be convincingly achieved by designing hierarchical cellular materials optimally even with moderate-quality constituents.

In recent years, topology optimization has become an efficient approach to fulfill this task. Successful applications are rapidly recognized for the purpose of tailoring effective properties of cellular materials. Among others, an inverse homogenization method was proposed by Sigmund [SIG 94, SIG 95], Sigmund and Torquato [SIG 97] as a material design procedure. The homogenization method that allows us to establish macroscopic effective properties of the heterogeneous medium in terms of microstructural variables was coupled with the SIMP model such that materials can be efficiently tailored to attain optimized microstructures or some prescribed elastic even extreme properties. Typical designs are shown in Figure 1.9. Similarly, a strain energy method was also developed by Zhang *et al.* [ZHA 07] to favor numerical implementation.

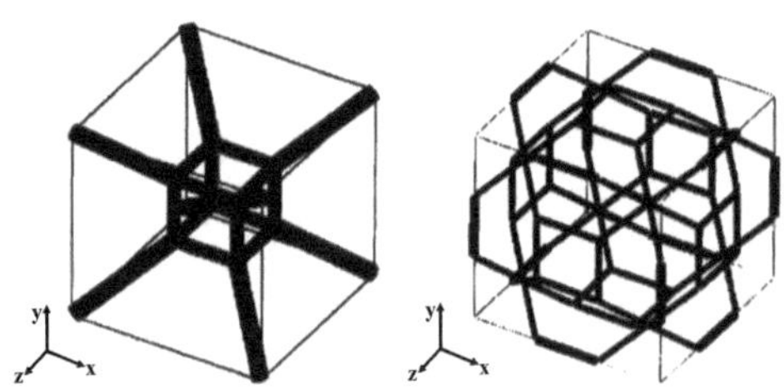


**Figure 1.9.** *Different 3D cells with prescribed Poisson's ratio [SIG 95]*

However, as the homogenization method is a two-scale asymptotic method based on the periodicity assumption of the microstructures, the predicted effective properties only depend upon the material microstructure, volume fractions and properties of constituents. Mathematically, under the periodicity assumption of the microstructures, the asymptotic expansion of each physical field, for example, the displacement of an arbitrary material point in a cellular elastic body is expressed as

$$u(x)=u^0(x,y)+\chi u^1(x,y)+\chi^2 u^2(x,y)+\ldots \qquad [1.9]$$

where $x$ is the macroscopic variable measured in the macroscale system ($X$) and varies slowly from unit cell to unit cell. $y= x/\chi$ is the periodic microscopic variable measured in the microscale system ($Y$) and varies quickly within each unit cell. Parameter $\chi$ is a small value representing the aspect ratio between the micro- and macroscale. Due to the complexity of analyzing the cellular structure illustrated in Figure 1.10 directly, unit cells of periodic microstructures are often modeled as a homogenized medium with effective elastic tensor. Based on the asymptotic expansion of equation [1.9], the elasticity equilibrium equation system defined over the unit cell can be expressed as:

$$\int_Y D_{ijpq} \frac{\partial u_p^{kl}}{\partial y_q} \frac{\partial \upsilon_i}{\partial y_j} dY = \int_Y D_{ijkl} \frac{\partial \upsilon_i}{\partial y_j} dY \qquad \forall \upsilon \in Y \qquad [1.10]$$

The effective elastic tensor of a periodic microstructure is then written as:

$$D_{ijkl}^{\mathrm{H}} = \frac{1}{|Y|} \int_Y \left( D_{ijkl} - D_{ijpq} \frac{\partial u_p^{kl}}{\partial y_q} \right) dY = \left(D_{ijkl}\right)^* - \left(\sigma_{ij}^{kl}\right)^* \qquad [1.11]$$

where $(D_{ijkl})^*$ denotes the average elastic tensor depending upon the material volume fractions of constituents as evaluated by the classical mixture rule. $(\sigma_{ij}^{kl})^*$ is the average stress tensor associated with the displacement vector $u^{kl}$ over the unit cell in load case $kl$ and represents a correction term reflecting the influence of the material microstructure of the unit cell. It is necessary to note that effective heat conductivity coefficients and thermal expansion coefficients of a microstructure can be evaluated in a similar way.

It should be mentioned that such a formulation is valid only asymptotically with $\chi$ being infinitesimally small and the obtained results preclude any scale-effect in the real structure even for stiffness design. In reality, this theoretical limit can never be reached. In other words, the homogenized descriptions are only valid when the size of the macrostructure is very large compared with the size of its microstructural heterogeneities. This situation was confirmed in the earlier study of buckling design by Bendsøe and Triantafyllidis [BEN 90]. Therefore, it is of great interest to formulate the integrated design problems with the retention of scale-effect for a real structure.

### 1.3.2. *Scale-effect of the material microstructure*

In contrast, strain energy minimization of cellular structures corresponds to a multiscale problem that is different from both the material design and pure structural design. The essential influence of underlying material microstructures upon macrostructure behavior requires that microstructures be designed to optimally match the loading and boundary conditions of the specific macrostructure. In this context, volume fractions of solid phases, the microstructure topology and the scale size of the material microstructure all have to be taken into account simultaneously.

Till now, Fujii *et al.* [FUJ 01] have studied the strain energy minimization of the macrostructure through topology optimization of material microstructures using the homogenization method. Rodrigues *et al.* [ROD 02] proposed a hierarchical computational procedure that integrates the global topology and local material design. As there may exist multiple solutions of microstructures that produce different local optima of effective elastic properties, Neves *et al.* [NEV 02] introduced the local buckling constraint to penalize the microstructure in the optimization procedure.

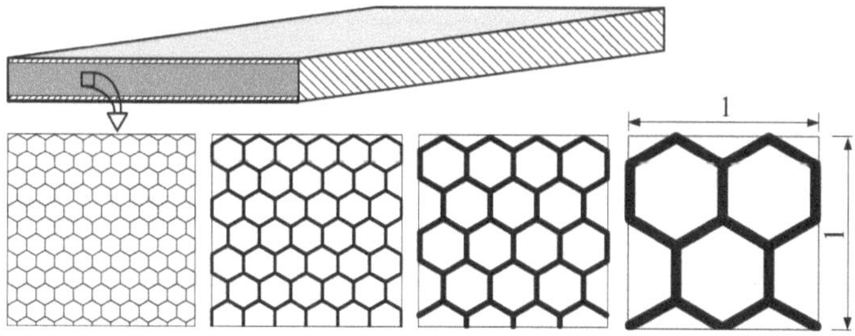


**Figure 1.10.** *Scale-effect of cellular core with identical microstructure and volume fraction*

In fact, one important issue in the simultaneous design of materials and structures is concerned with the scale-effect of the microstructure, as shown in Figure 1.10. This mechanism was confirmed from both micromechanics theories and experiments [SUT 99, PEC 99]. Recently, Tantikom *et al.* [TAN 05] pointed out this phenomenon experimentally. As shown in Figure 1.11, the specimen with tubular core results in different nominal stress-strain curves when the number of cellular layers changes. Curves become indifferent whenever the number of cellular layers is large enough. This mechanism is the so-called scale-effect as discussed by Burgueno *et al.* [BUR 05] and Dai and Zhang [DAI 08]. Therefore, the scale-effect should be taken into account together with the morphology of the microstructure in a comprehensive way in the design optimization.

However, as the homogenization method leads to the same effective elastic tensor in all above cases of microstructures, its utilization is reasonable only when the number of cells involved in the core is large enough. On the contrary, when the panel has only a few cellular layers, the scale-effect becomes important in the design so that the predicted effective properties are unable to account for the real state of the material microstructure.

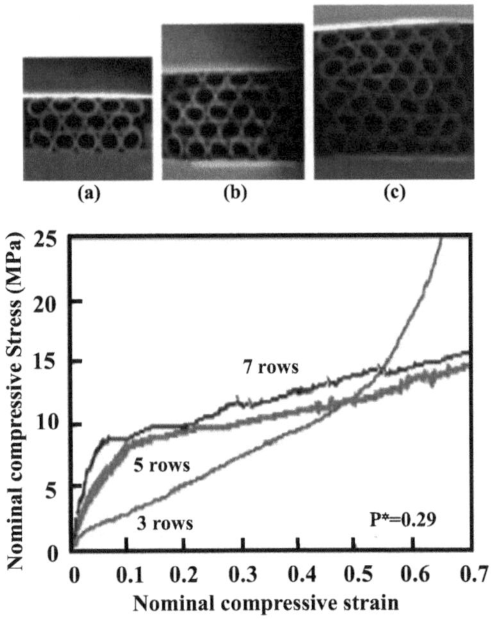


**Figure 1.11.** *Specimens with three, five and seven layers and in-plane nominal compressive stress-strain curves of pure copper assemblies with different numbers of rows [TAN 05]*

Here, an integrated optimization approach [ZHA 06] is presented for the strain energy minimization of two-dimensional (2D) layered structures. The scale-effect of the microstructures upon the topologically optimized design is highlighted. Without loss of generality, we consider the periodicity of the cellular microstructure in a given design domain, as shown in Figure 1.12. The scale-related design optimization may be interpreted as how to determine the number of unit cells as well as the involved microstructures. In fact, such a scale-related problem reflects the intrinsic dependence between the material and structure. With the given material volume fraction, increasing $m$, i.e. the number of the unit cells, means reducing the size of each microstructure and the small parameter $\chi$ defined as the ratio between the microscale of the material and the macroscale of the structure. In the ultimate case of an infinite number of unit cells, the scale-effect will become ignorable and the design result converges to the asymptotic solution of the homogenization method. Alternatively, the limit case of involving only one unit cell corresponds to topology optimization of the pure macrostructure.

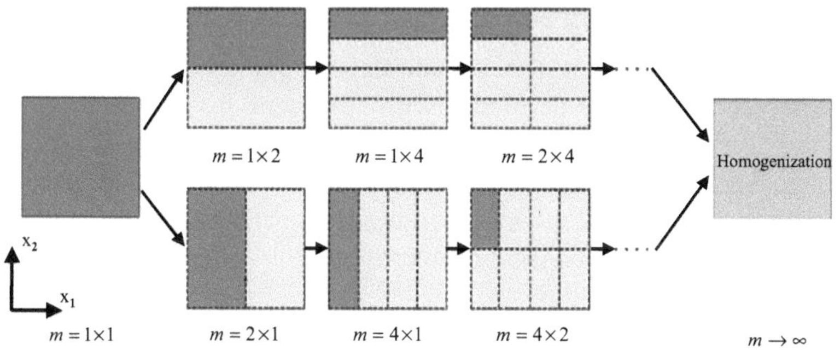


**Figure 1.12.** *A two-dimensional domain with different numbers of unit cells*

### 1.3.3. *Scale-related topology optimization*

For illustration, a cellular solid in the macroscale ($X$) with known boundary conditions and external forces is shown in Figure 1.13. The optimized solution is one that uses a spatial distribution of heterogeneous cellular materials characterized by different microstructures at different locations of the design domain. Thus, materials will be optimally distributed to match the loaded regions. To do this, the design procedure is partitioned into two steps: macroscale layout optimization and refined microstructure optimization.

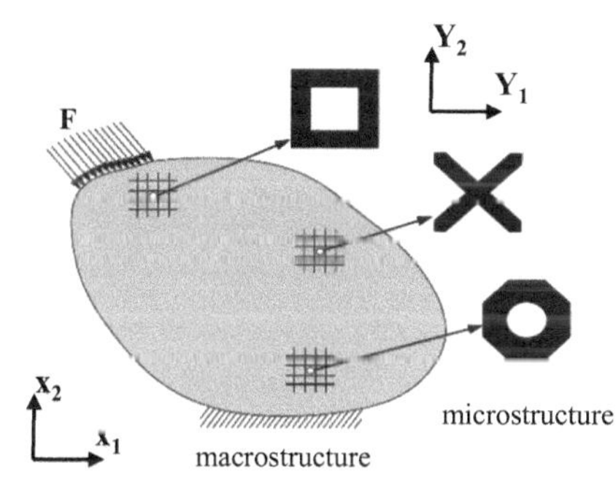


**Figure 1.13.** *Illustration of a cellular domain with different material microstructures*

First, the macrostructure is considered as a homogenized body and discretized into a finite element model. The overall behavior can be determined by solving the corresponding FE equilibrium equation system. For each element, suppose that the element stiffness matrix depends upon the element pseudo-density variable linearly. This corresponds to the SIMP model with exponent $p = 1$ for a smooth variation of the optimized pseudo-density variables over the design domain. Suppose the structural strain energy is minimized subjected to the volume constraint. The formulation corresponds to exactly a traditional sizing optimization problem.

$$\begin{aligned}
&(\text{Macroscale for global structure})\\
&\text{find:} \quad \boldsymbol{\eta}^{(\mathrm{A})} = \left\{\eta_i^{(\mathrm{A})}\right\} \quad i = 1, 2, \ldots, n_{\mathrm{e}}^{(\mathrm{A})}\\
&\text{min:} \quad C = \frac{1}{2}\mathbf{F}^{\mathrm{T}}\mathbf{U}\\
&\text{s.t.} \quad V \le V_{\mathrm{U}}\\
&\qquad\; 0 < \eta_i^{(\mathrm{A})} \le 1
\end{aligned} \qquad [1.12]$$

with

$$\begin{aligned}
&(\text{Macroscale for global structure})\\
&E_i = \eta_i^{(\mathrm{A})} E_{i0}
\end{aligned} \qquad [1.13]$$

Once the above mentioned macroscale optimization problem is solved, a global distribution of pseudo-densities is obtained over the macrostructure. If values of pseudo-densities attain nearly 0 or 1, it means that the corresponding element is a void or solid one. Otherwise, the element has a cellular microstructure with intermediate pseudo-density and needs to be further refined at the step of the microscale design. To this end, all regions of intermediate pseudo-density values are first identified and grouped into subdomains and then topology optimization is carried out to determine microstructures and their sizes following design specifications and available manufacturing capabilities. Meanwhile, each unit cell considered as a subdomain can be allowed to have its proper microstructure. Alternatively, we can also consider elements of the same pseudo-density values as a group of subdomains to hold the identical microstructure distributed periodically. Here, all subdomains are considered to be unit cells of identical size, which are further discretized into $n_e^{(B)}$ finite elements with their own pseudo-density variables. For unit cell $i$, the problem of microstructure topology optimization can be similarly defined as

$$\begin{aligned}
&(\text{Microscale for unit cell } i) \\
&\text{find:} \quad \boldsymbol{\eta}_i^{(B)} = \left\{\eta_{i,j}^{(B)}\right\} \quad j = 1,2,\ldots,n_e^{(B)} \\
&\max: \quad C_i = \frac{1}{2}\mathbf{U}_i^{T}\mathbf{K}\mathbf{U}_i \\
&\text{s.t.} \quad V_i \le V_{i0}\eta_i^{(A)} \\
&\qquad\quad 0 < \eta_{i,j}^{(B)} \le 1
\end{aligned} \qquad [1.14]$$

where the microstructure is optimized in terms of the pseudo-densities for each unit cell in microscale. $V_{i0}$ is the volume of the unit cell $i$ with full material. The strain energy related to the objective function is also calculated locally in the microstructure.

It should be noted that: (1) the strain energy in microscale is calculated according to the known nodal displacement vector in macroscale; (2) as the displacement field is known, the stiffness design now corresponds to a maximization problem of strain energy and (3) the upper bound of the volume constraint for each unite cell is also inherited from the macroscale design variable $\eta_i^{(A)}$.

To figure out the optimal topology, pseudo-density variables in microscale are penalized by the SIMP law, e.g. $p = 4$ in the current design procedure. To perform a sensitivity analysis of the objective function, two approaches are set up as follows.

#### 1.3.3.1. *The scale-related approach*

For element $j$ included in the unit cell $i$, assume that the dependence of element stiffness matrix upon its pseudo-density variable obeys the SIMP law. Similarly to a topology optimization in macroscale, the sensitivity of the objective function with respect to each pseudo-density variable in microscale can be directly calculated as follows:

$$\frac{\partial C_i}{\partial \eta_{i,j}^{(\mathrm{B})}} = \frac{1}{2}\mathbf{U}_i^{\mathrm{T}} \frac{\partial \mathbf{K}_i}{\partial \eta_{i,j}^{(\mathrm{B})}} \mathbf{U}_i = \frac{1}{2}\mathbf{U}_i^{\mathrm{T}} \frac{\partial \mathbf{K}_{i,j}}{\partial \eta_{i,j}^{(\mathrm{B})}} \mathbf{U}_i = \frac{1}{2}\frac{p}{\eta_{i,j}^{(\mathrm{B})}} C_{i,j} \qquad [1.15]$$

where $\mathbf{K}_{i,j}$ and $C_{i,j}$ denote the stiffness matrix and strain energy of element $j$ in unit cell $i$.

#### 1.3.3.2. *The homogenization approach*

When the unit cell is specified to have a small enough size the homogenization method can be applied to evaluate the effective elastic matrix $\mathbf{D}^{\mathrm{H}}$. In such a way, the stiffness matrix of one such element depends upon the microscale pseudo-densities as follows:

$$\mathbf{K}_i = \int \mathbf{B}_i \mathbf{D}_i^{\mathrm{H}} \left(\boldsymbol{\eta}_i^{(\mathrm{B})}\right) \mathbf{B}_i \mathrm{d}\Omega \qquad [1.16]$$

where $\mathbf{B}_i$ is the strain-displacement matrix or the geometry matrix. Based on the finite element discretization of unit cell $i$, the effective elastic matrix is evaluated according to the homogenization method in its discrete form:

$$\mathbf{D}_i^{\mathrm{H}} = \frac{1}{V_{i0}} \sum_j \int \left( \mathbf{D}_{i,j}\left(\eta_{i,j}^{(\mathrm{B})}\right) - \mathbf{D}_{i,j}\left(\eta_{i,j}^{(\mathrm{B})}\right) \mathbf{B}_{i,j} \mathbf{U}_{i,j}\left(\boldsymbol{\eta}_i^{(\mathrm{B})}\right)\right) \mathrm{d}\Omega \qquad [1.17]$$

$\mathbf{D}_{i,j}$ and $\mathbf{U}_{i,j}$ are the elastic matrix and nodal displacement vector of element $j$ involved in unit cell $i$.

According to the SIMP interpolation model, the derivative of the effective elastic matrix can be written as:

$$\begin{aligned}\frac{\partial \mathbf{D}_i^{\mathrm{H}}}{\partial \eta_{i,j}^{(\mathrm{B})}} &= \frac{1}{V_{i0}}\left(\frac{p}{\eta_{i,j}^{(\mathrm{B})}}\mathbf{D}_{i,j}V_{i,j0} - \frac{p}{\eta_{i,j}^{(\mathrm{B})}}\int \mathbf{D}_{i,j}\mathbf{B}_{i,j}\mathbf{U}_{i,j}\left(\boldsymbol{\eta}_i^{(\mathrm{B})}\right)\mathrm{d}\Omega - \sum_j \int \mathbf{D}_{i,j}\mathbf{B}_{i,j}\frac{\partial \mathbf{U}_{i,j}}{\partial \eta_{i,j}^{(\mathrm{B})}}\mathrm{d}\Omega\right) \\ &= \frac{1}{V_{i0}}\left(\frac{p}{\eta_{i,j}^{(\mathrm{B})}}\mathbf{D}_{i,j}V_{i,j0} - \frac{p}{\eta_{i,j}^{(\mathrm{B})}}\int \boldsymbol{\sigma}_{i,j}\left(\boldsymbol{\eta}_i^{(\mathrm{B})}\right)\mathrm{d}\Omega - \sum_j \int \mathbf{D}_{i,j}\mathbf{B}_{i,j}\frac{\partial \mathbf{U}_{i,j}}{\partial \eta_{i,j}^{(\mathrm{B})}}\mathrm{d}\Omega\right)\end{aligned} \quad [1.18]$$

Now, based on these intermediate results, the chain rule will be finally applied for the differentiation of the objective function such that

$$\frac{\partial C_i\left(\boldsymbol{\eta}_i^{(\mathrm{B})}\right)}{\partial \eta_{i,j}^{(\mathrm{B})}} = \frac{1}{2}\mathbf{U}_i^{\mathrm{T}}\frac{\partial \mathbf{K}_i}{\partial \boldsymbol{\eta}_{i,j}^{(\mathrm{B})}}\mathbf{U}_i = \frac{1}{2}\mathbf{U}_i^{\mathrm{T}}\left(\int \mathbf{B}_i^T \frac{\partial \mathbf{D}_i^{\mathrm{H}}}{\partial \boldsymbol{\eta}_{i,j}^{(\mathrm{B})}}\mathbf{B}_i \mathrm{d}\Omega\right)\mathbf{U}_i \quad [1.19]$$

Obviously, the scale-related approach is a simple and convenient formulation to obtain the design sensitivities needed for topology optimization when compared with the homogenization approach.

### 1.3.4. *Numerical examples*

Consider now a 2D rectangular domain of plane stress state, as shown in Figure 1.14. Assume that the design domain has a dimension of 32 × 20 m and a thickness of 1 m. The panel is loaded vertically with 100 N/m. Young's modulus and Poisson's ratio of the material are 1,000 Pa and 0.3, respectively. In this problem, a volume fraction of 60% is used for the solid material in the design domain and the dimension of the unit cell, i.e. representative volume element (RVE) is noted by $l \times h$. For finite element modeling, the problem will be ill-conditioned and become singular if solid elements are eliminated along the right edge where the vertical load is applied directly. To avoid this, a small non-designable elastic portion will be added artificially along one such edge to transfer the applied load.

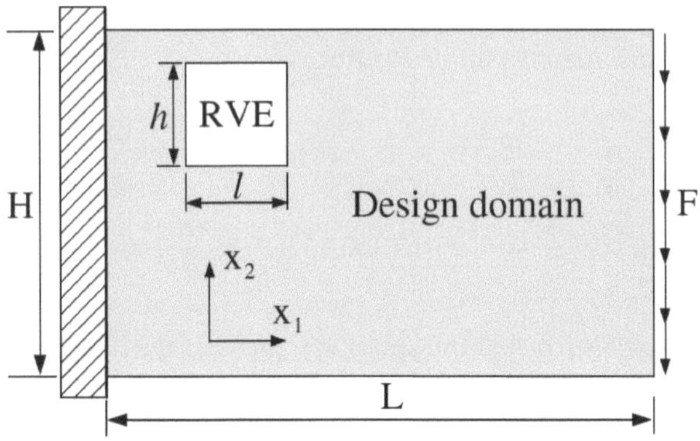


**Figure 1.14.** *2D rectangular design domain*

The 2D panel is macroscopically discretized into a mesh of 20 × 32 finite elements. Assume that elements in each horizontal layer have the same value of pseudo-density variables that can be realized in the way of design variable linking technique. Consequently, 20 independent design variables exist. The optimized distribution of pseudo-density variables in macroscale is shown in Figure 1.15. This is exactly equivalent to a sizing problem. A symmetric distribution is obtained with values of pseudo-density variables being (1.0, 0.72, 0.46, 0.41, 0.46, 0.72 and 1.0). As expected, such a graded distribution is the stiffest to prevent the bending deformation from the engineering viewpoint.

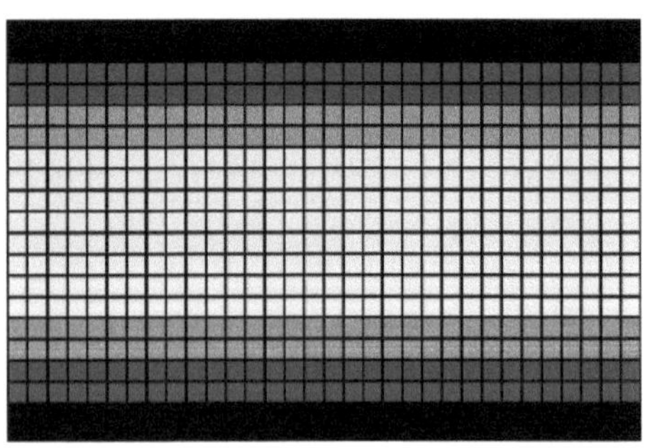

**Figure 1.15.** *Macroscale design for material layout*

According to the result given in Figure 1.15, unit cells in each horizontal layer are assumed to have an identical microstructure and eight independent subproblems will be solved. Based on the scale-related approach, the optimized solution related to 16 × 8 unit cells is given in Figure 1.16(a) with the corresponding strain energy denoted by $C_{16\times8}$.

Comparatively, if the homogenization method is used, the effective elastic tensor is evaluated for each macroelement meshed with 32 × 20 microelements. The optimized solution denoted by $C_{\infty}$ is shown in Figure 1.16(b). We can see that microstructures change from layer to layer in order that load-bearing capacities of materials are fully explored to resist the bending force. In particular, a lozenge microstructure is obtained along the neutral axis to resist the maximum shear stress. Note that although values of the structural strain energy obtained by both approaches are very close, a difference still exists between microstructures.

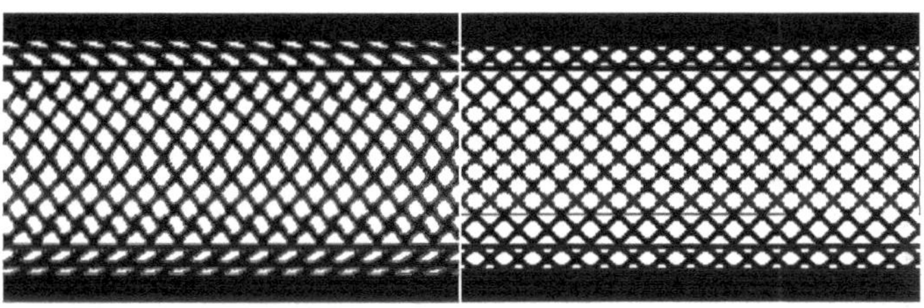

a) $C_{16\times8}$ =172 019.7 b) $C_{\infty}$ =172 171.8

**Figure 1.16.** *Comparison of scale-related design and homogenization-based design*

To further reveal the scale-effect, the influence of the number of unit cells upon the optimized designs of microstructures is studied now. Suppose that the inner core has a uniform value of 0.5 for pseudo-density variables and all unit cells in the inner core have a common microstructure. This corresponds to the same volume fraction (60%) over the whole structure but a volume fraction of 50% over the core. Figure 1.17 shows that a variety of design solutions are obtained for square unit cells of different scales. The comparison indicates that a successive size diminution of the unit cell reduces the design space and increases the number of design variable linking constraints so that the structural strain energy increases and converges to the solution of the homogenization method. This observation indicates that both material and structural designs can be unified from the viewpoint of the relative scale and the homogenization-based design is a limit solution case with infinite unit cells.

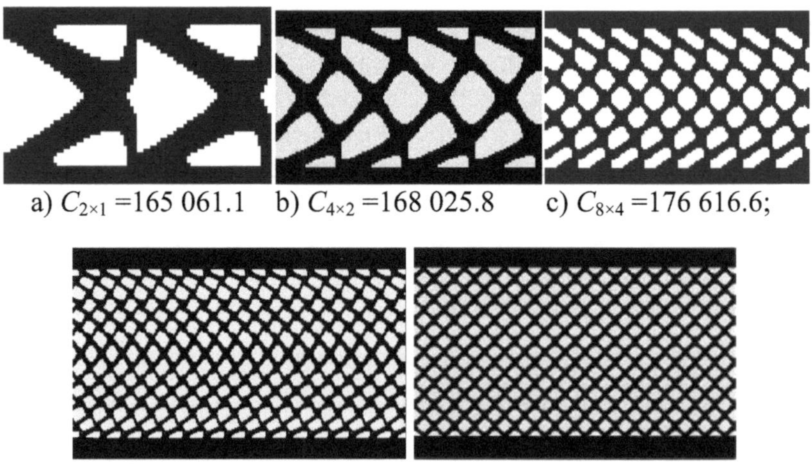

a) $C_{2\times1}$ =165 061.1 b) $C_{4\times2}$ =168 025.8 c) $C_{8\times4}$ =176 616.6;

d) $C_{16\times8}$ =181 094.94 e) $C_{\infty}$ =186 282.8

**Figure 1.17.** *Scale-effect of the square RVE on the core design result*

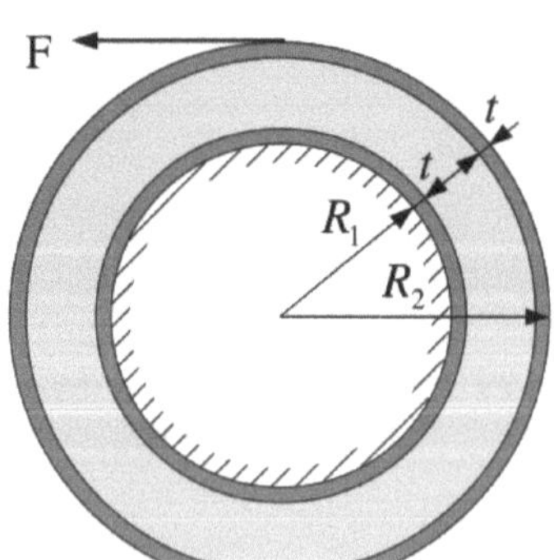


**Figure 1.18.** *Cellular core design optimization*

The proposed approach [ZHA 06] can also be extended to the cellular core design of circular structures that find wide applications in the aerospace industry. The test is about a circular membrane with inner and outer skins as shown in Figure 1.18. The inner hole and outer contour have radii of 60 and 120 mm, respectively. Two cases of different skin thicknesses of 6 and 1 mm are investigated. Fixations will be imposed along the inner hole. Under the point-wise tangential loading of 1,600 N applied on the outer contour, the cellular core limited to a volume fraction of 40% needs to be designed for a maximum rigidity. To do this, the structure is partitioned into 16 representative sectors, i.e. unit cells that hold the cyclic symmetry.

By performing the refined design procedure using the scale-related approach, optimal design patterns are illustrated in Figure 1.19. It has been found that stiffeners have a layout nearly perpendicular to each other like the classical solution of the Michell truss structure. Besides, both solutions are very similar except for small holes near the outer contour when the skin thickness is 6 mm. These holes are completely filled when the skin thickness is 10 mm.

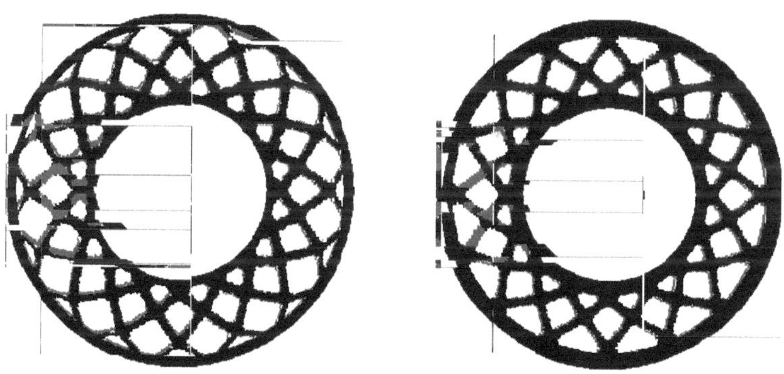

**Figure 1.19.** *Optimized designs of cellular core with outer skin thicknesses of 6 and 10 mm*

By keeping the skin thickness of 6 mm unchanged, two more examples are tested by partitioning the structure into 8 and 24 unit cells. The optimized designs are compared in Figure 1.20.

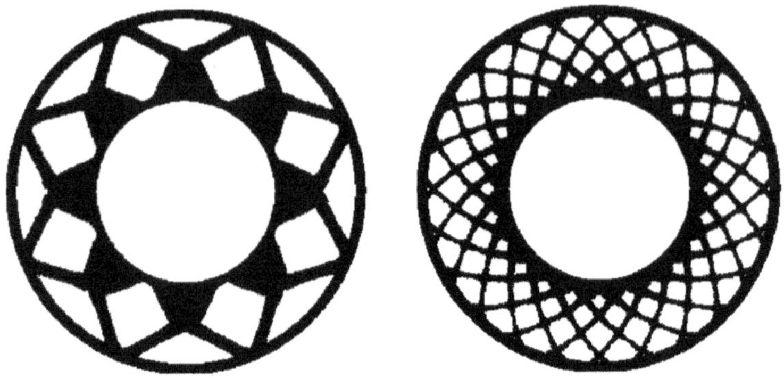

**Figure 1.20.** *Optimized designs of cellular core with 8 and 24 unit cells*

Apart from the above work, Yan *et al.* [YAN 08] and Liu *et al.* [LIU 08] proposed a concurrent optimization procedure to design uniform cellular materials as well as macrostructures. Figure 1.21 presents one typical design where a mechanical load and a temperature rise are applied simultaneously.

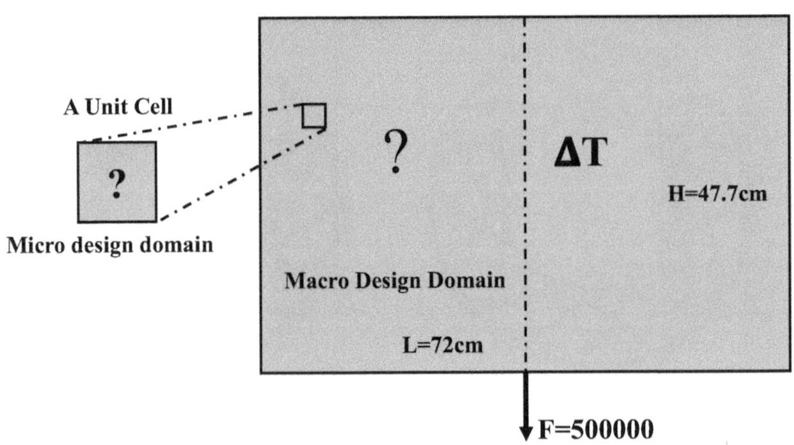


**Figure 1.21.** *Design problem with the simultaneous presence of a mechanical load and a temperature rise*

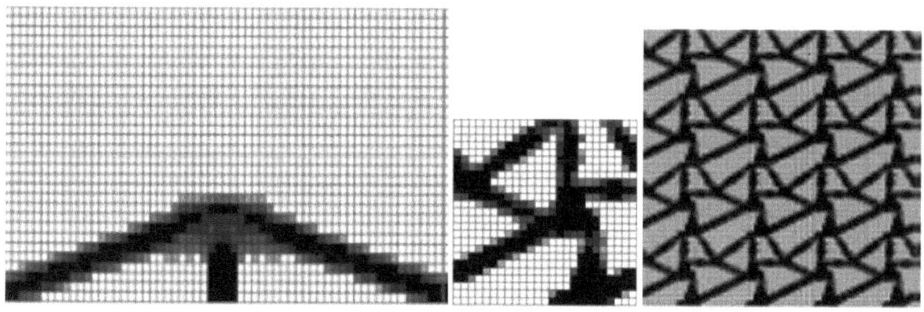

**Figure 1.22.** *Simultaneous topology optimization of microstructures and macrostructures [YAN 08]*

Note that Rodrigues *et al.* [ROD 02] and Coelho *et al.* [COE 08] carried out topology optimization of macrostructures and microstructures in a hierarchical way without accounting for scale-effect. An optimized design is shown in Figure 1.23. Different microstructures are obtained in different locations to match the local material properties calculated from the global material distribution and loading conditions.

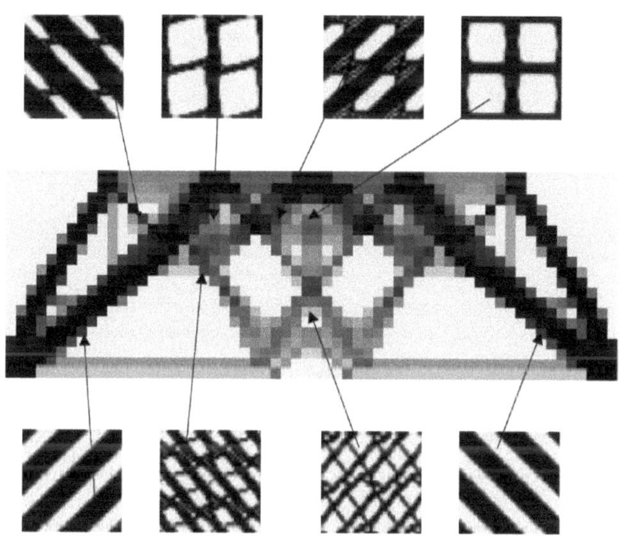

**Figure 1.23.** *Hierarchical optimization of macrostructures and microstructures*

## 1.4. Conclusions

Standard material layout design is the basic form of topology optimization. By defining the design variables related to the finite element mesh and assuming the constraint to material volume, topology optimization of macrostructure, microstructure and their integration are made possible.

Microstructure design or material design is mostly formulated based on the homogenization method to find proper microstructure patterns satisfying the prescribed material properties. It is, however, scale-independent. With the introduction of the subdomain concept, we investigate the scale-effect of the microstructure upon the optimized topology. The integrated design of materials and structures is introduced as a two-level design methodology combining the macroscale layout optimization with the refined design of microstructures.

The scale-effect of the microstructure upon the optimized topology solution is highlighted. In limit cases, the design becomes a pure topology optimization problem of the macrostructure or a pure topology optimization problem of the microstructure that is asymptotically equivalent to the homogenization solution. Therefore, it concludes that designs of material microstructures and macrostructures are relative solutions depending upon the scale. Likewise, designs of periodical cellular structures can also be extended to circular structures of cyclic symmetry.

# 2

# Low-Density Areas in Topology Optimization

## 2.1. Localized mode in low-density areas

Localized mode often appears in topology optimization for maximizing the natural frequencies or buckling loads [NEV 95, PED 00, ZHO 04]. It means that the vibration or buckling takes place only in the low-density areas related to void elements that should not physically have a mechanical effect. Consequently, structural responses and sensitivities are incorrectly calculated which misleads the optimization process. This phenomenon is actually recognized as a type of numerical singularity due to the improperly defined material properties for the void elements, especially when the popular SIMP model is directly used.

A simple example of a cantilevered beam is illustrated in Figure 2.1. The cross-section of the beam is a square with a size of 0.05 m × 0.05 m. Suppose that the beam is discretized into two elements. Only nodes $A$ and $B$ are free and 4 degrees of freedom exist altogether, i.e. the vertical displacements and the rotations of the two nodes ($v_A, \theta_A, v_B, \theta_B$). Based on the SIMP interpolation model, pseudo-density variables are penalized in different orders and related to material properties in the following form:

Element 1: elastic modulus $E_1 = \eta_1^3 \times 10^6$ Pa, density $\rho_1 = \eta_1 \times 100$ kg/m$^3$, Poisson's ratio 0.3;

Element 2: elastic modulus $E_2 = \eta_2^3 \times 10^6$ Pa, density $\rho_2 = \eta_2 \times 100$ kg/m$^3$, Poisson's ratio 0.3;

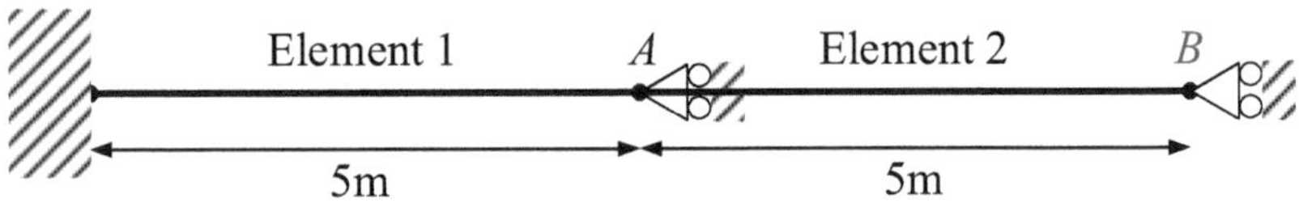


**Figure 2.1.** *A cantilevered beam with two elements*

Theoretically, the stiffness and mass matrices of element $i$ are expressed as:

$$\mathbf{K}_i = \begin{bmatrix} \frac{12E_iI}{L^3} & \frac{6E_iI}{L^2} & -\frac{12E_iI}{L^3} & \frac{6E_iI}{L^2} \\ \frac{6E_iI}{L^2} & \frac{4E_iI}{L} & -\frac{6E_iI}{L^2} & \frac{2E_iI}{L} \\ -\frac{12E_iI}{L^3} & -\frac{6E_iI}{L^2} & \frac{12E_iI}{L^3} & -\frac{6E_iI}{L^2} \\ \frac{6E_iI}{L^2} & \frac{2E_iI}{L} & -\frac{6E_iI}{L^2} & \frac{4E_iI}{L} \end{bmatrix} \quad [2.1]$$

$$\mathbf{M}_i = \frac{\rho_i AL}{420} \begin{bmatrix} 156 & 22L & 54 & -13L \\ 22L & 4L^2 & 13L^2 & -3L^2 \\ 54 & 13L^2 & 156 & -22L \\ -13L & -3L^2 & -22L & 4L^2 \end{bmatrix} \quad [2.2]$$

where $A$ and $I$ are the area and moment of inertia of the beam cross-section. $L$ is the length of one beam element.

By assembling the element matrices and removing the DOFs fixed by the boundary conditions, the global stiffness and mass matrices are expressed as:

$$\mathbf{K}=\begin{bmatrix} \frac{12I}{L^3}(E_1+E_2) & \frac{6I}{L^2}(E_1-E_2) & -\frac{12E_2I}{L^3} & \frac{6E_2I}{L^2} \\ \frac{6I}{L^2}(E_1-E_2) & \frac{4I}{L}(E_1+E_2) & -\frac{6E_2I}{L^2} & \frac{2E_2I}{L} \\ -\frac{12E_2I}{L^3} & -\frac{6E_2I}{L^2} & \frac{12E_2I}{L^3} & -\frac{6E_2I}{L^2} \\ \frac{6E_2I}{L^2} & \frac{2E_2I}{L} & -\frac{6E_2I}{L^2} & \frac{4E_2I}{L} \end{bmatrix} \quad [2.3]$$

$$\mathbf{M}=\frac{1}{420}\begin{bmatrix} 156AL(\rho_1+\rho_2) & -22AL^2(\rho_1-\rho_2) & 54\rho_2AL & -13\rho_2AL^2 \\ 22AL^2(\rho_1-\rho_2) & 4AL^3(\rho_1+\rho_1) & 13\rho_2AL^3 & -3\rho_2AL^3 \\ 54\rho_2AL & 13\rho_2AL^3 & 156\rho_2AL & -22\rho_2AL^2 \\ -13\rho_2AL^2 & -3\rho_2AL^3 & -22\rho_2AL^2 & 4\rho_2AL^3 \end{bmatrix} \quad [2.4]$$

First, suppose elements 1 and 2 are solid with pseudo-density variables $\eta_1 = 1$ and $\eta_2 = 1$. The assembled global stiffness and mass matrices correspond to

$$\mathbf{K}=\begin{bmatrix} 0.1000 & 0 & -0.0500 & 0.1250 \\ 0 & 0.8333 & -0.1250 & 0.2083 \\ -0.0500 & -0.1250 & 0.0500 & -0.1250 \\ 0.1250 & 0.2083 & -0.1250 & 0.4167 \end{bmatrix} \quad [2.5]$$

$$\mathbf{M}=\begin{bmatrix} 0.9286 & 0 & 0.1607 & -0.1935 \\ 0 & 0.5952 & 0.1935 & -0.2232 \\ 0.1607 & 0.1935 & 0.4643 & -0.3274 \\ -0.1935 & -0.2232 & -0.3274 & 0.2976 \end{bmatrix} \quad [2.6]$$

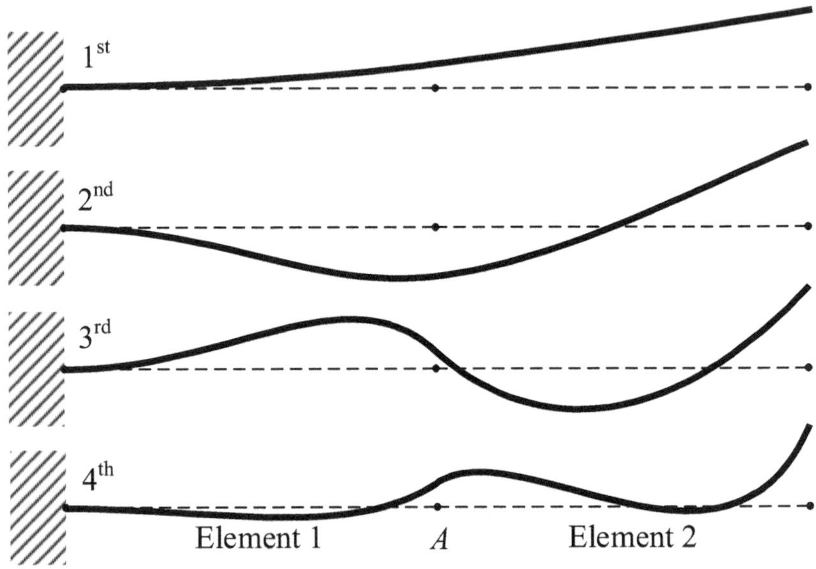


**Figure 2.2.** *Four mode shapes of the cantilevered beam*

By solving the eigenequation, the four eigenvalues and corresponding eigenvectors are obtained as:

$$\left[\omega_1^2, \omega_2^2, \omega_3^2, \omega_4^2\right] = \left[0.0026 \quad 0.1029 \quad 1.1768 \quad 9.9134\right] \quad [2.7]$$

$$\left[v_A, \theta_A, v_B, \theta_B\right]^{\mathrm{T}} = \begin{bmatrix} 0.3395 & -0.7194 & 0.1017 & 0.2531 \\ 0.1163 & 0.0433 & -0.7647 & 0.5204 \\ 1 & 1 & 1 & 1 \\ 0.1377 & 0.4799 & 0.9644 & 1.9329 \end{bmatrix} \quad [2.8]$$

The first- to fourth-order mode shapes are shown in Figure 2.2. Nothing abnormal is observed in the obtained natural frequencies and mode shapes. Notice that the eigenvectors are normalized by assuming the vertical displacement of node $B$ to be 1.

Now, the material property of element 2 is perturbed by setting $\eta_2 = 0.01$ to represent a void element of low-density in topology optimization. The global stiffness and mass matrices become

$$\mathbf{K} = \begin{bmatrix} 0.0500 & -0.1250 & -5.0\times10^{-8} & 1.3\times10^{-7} \\ -0.1250 & 0.4167 & -1.3\times10^{-7} & 2.1\times10^{-7} \\ -5.0\times10^{-8} & -1.3\times10^{-7} & 5.0\times10^{-8} & -1.3\times10^{-7} \\ 1.3\times10^{-7} & 2.1\times10^{-7} & -1.3\times10^{-7} & 4.2\times10^{-7} \end{bmatrix} \quad [2.9]$$

$$\mathbf{M} = \begin{bmatrix} 0.4689 & -0.3241 & 0.0016 & -0.0019 \\ -0.3241 & 0.3006 & 0.0019 & -0.0022 \\ 0.0016 & 0.0019 & 0.0046 & -0.0033 \\ -0.0019 & -0.0022 & -0.0033 & 0.0030 \end{bmatrix} \quad [2.10]$$

Notice that in the global stiffness matrix, the absolute values of terms related to element 2 decrease much faster than those in the mass matrix because of the power-law penalization. After solving the eigenequation, eigenvalues and eigenvectors are obtained as:

$$\left[\omega_1^2, \omega_2^2, \omega_3^2, \omega_4^2\right] = \left[4.2\times10^{-6} \quad 0.0004 \quad 0.0411 \quad 3.7938\right] \quad [2.11]$$

$$\left[v_A, \theta_A, v_B, \theta_B\right]^{\mathrm{T}} = \begin{bmatrix} 3.31\times10^{-6} & -7.29\times10^{-5} & 1.5345 & 0.7839 \\ 1.14\times10^{-6} & -2.37\times10^{-5} & 0.4234 & 1.2158 \\ 1 & 1 & 1 & 1 \\ 0.2755 & 1.5245 & 2.4250 & 2.5214 \end{bmatrix} \quad [2.12]$$

The first and second eigenvalues are greatly reduced because of the significant difference between the stiffness and mass matrices. Specifically, in the first and second eigenvectors, node $A$ has very small displacements and nearly stops vibrating. This indicates that only element 2 vibrates. Figure 2.3 shows that the vibrations act as localized modes only in the low-density element at the first- and second-order vibrations. The reason is that the material interpolation model leads to the mismatch between the stiffness and mass matrices.

Therefore, avoiding the localized mode is a serious problem to be dealt with in practice. In order to have a detailed view of the localized mode, the material property of element 2 is weakened step-by-step from $\eta_2 = 1$ to $\eta_2 = 0.01$. The modal analysis is carried out correspondingly to show the variations of the eigenvalues and mode shapes.

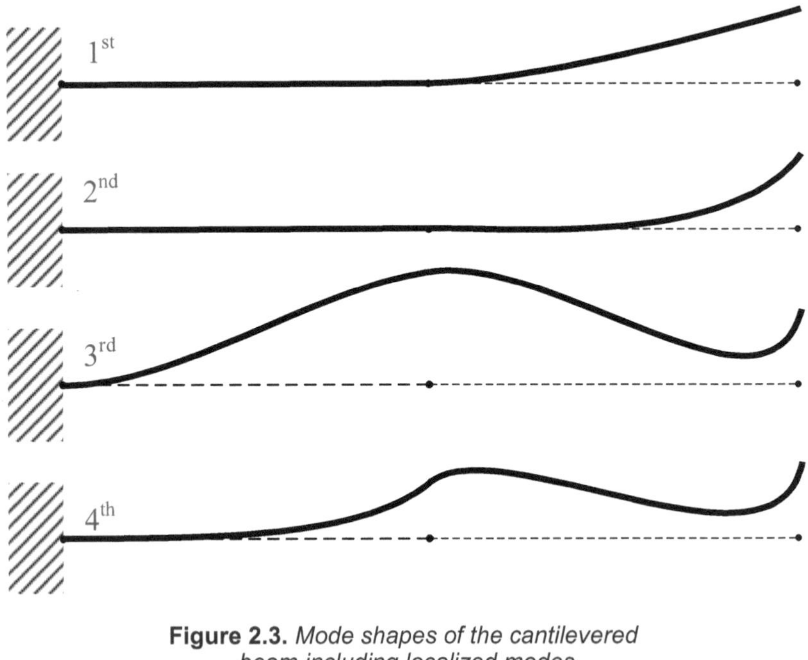


**Figure 2.3.** *Mode shapes of the cantilevered beam including localized modes*

Since the localized modes appear in the first- and second-order vibrations, only these two modes are analyzed. Figures 2.4 and 2.5 show the variations of the two eigenvectors and eigenvalues versus $\eta_2$.

Figure 2.4 indicates that the first-order vibration of element 1 decreases monotonically while element 2 tends toward void. Figure 2.5 indicates that the localized mode occurs continuously.

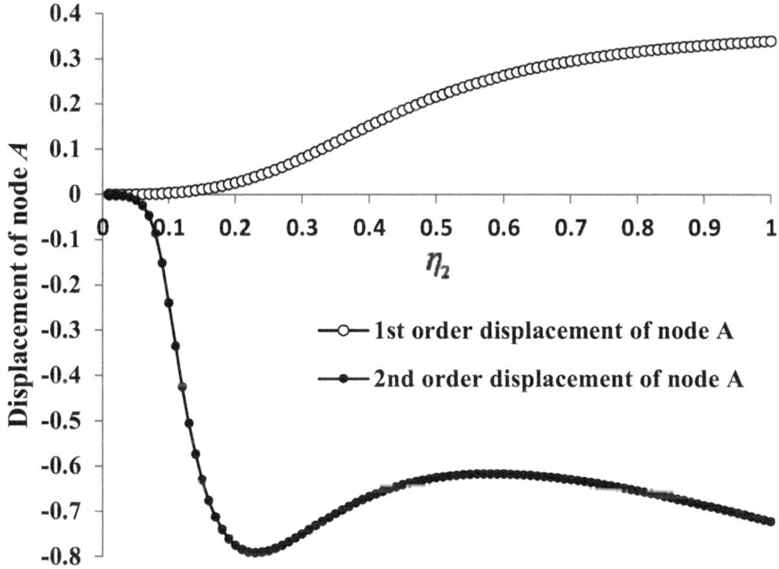


**Figure 2.4.** *Displacements of node A versus $\eta_2$*

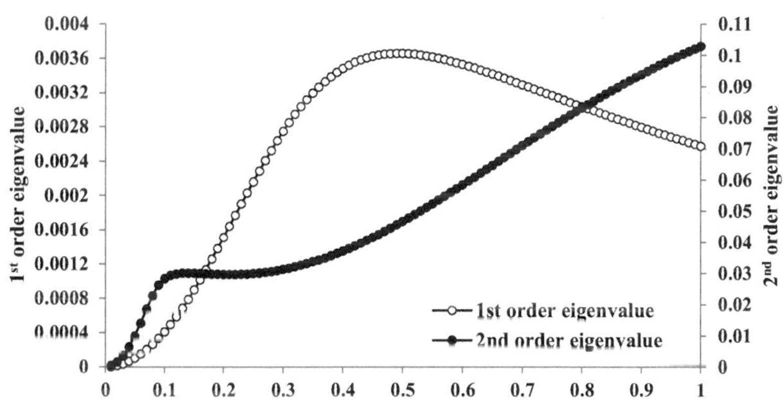


**Figure 2.5.** *First and second eigenvalues versus $\eta_2$*

It is obvious that no clear bound can be defined for the occurrence of localized mode. When the elements in the low-density area are compliant enough, the solid elements stop their vibrations and act as an approximately rigid part which can be removed from the stiffness and mass matrices. In this case, the eigenequation can be simplified as:

$$\left(\mathbf{K}_2 - \omega^2\mathbf{M}_2\right)\mathbf{U} = 0 \qquad [2.13]$$

so that

$$\begin{aligned}\omega^2\mathbf{U} &= \mathbf{M}_2^{-1}\mathbf{K}_2\mathbf{U} = \left(\eta_2\mathbf{M}_{2,0}\right)^{-1}\eta_2^3\mathbf{K}_{2,0}\mathbf{U} \\ &= \eta_2^2\mathbf{M}_{2,0}{}^{-1}\mathbf{K}_{2,0}\mathbf{U}\end{aligned} \qquad [2.14]$$

where $\mathbf{K}_2$ and $\mathbf{M}_2$ are the stiffness matrices of element 2, and $\mathbf{K}_{2,0}$ and $\mathbf{M}_{2,0}$ are the corresponding stiffness and mass matrices at the solid element state. Equation [2.14] shows that the first two eigenvalues related to the localized modes depend upon $\eta_2$ in the parabolic way. Figure 2.6 compares these variations with the solutions before simplification.

Curves related to first and second eigenvalues start to coincide with the simplified parabola at around $\eta_2 = 0.14$ and $\eta_2 = 0.06$, respectively. These coincidences are also confirmed by the displacement variations of the mode shape in Figure 2.4 where node *A* stops vibrating around these critical values.

According to the above analysis, the lower bound of the pseudo-density variable in the SIMP model must be greater than these critical values to ensure the threshold stiffness of the low-density areas. Although a lower penalty factor can be used in SIMP to moderate the mismatch between the stiffness and mass matrices, this will unfortunately weaken the key penalty effect of the SIMP model in generating clear black and white structural patterns. In conclusion, the standard SIMP interpolation model with the power-law penalization cannot directly be used in topology optimization of eigenvalue maximization.

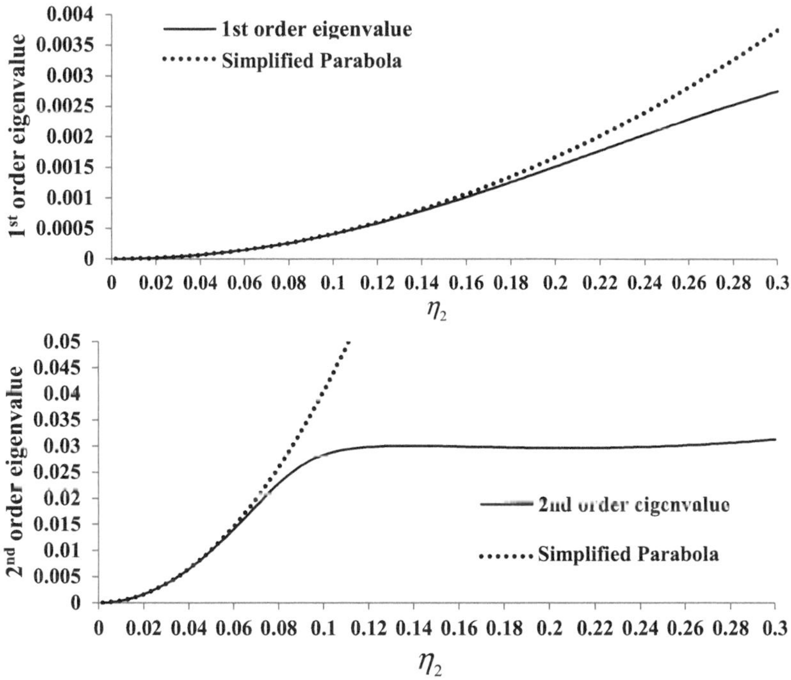


**Figure 2.6.** *Comparisons between real eigenvalues and simplified parabola*

A test of maximizing the fundamental frequency is further carried out to show how the localized modes influence the topological design. As illustrated in Figure 2.7, a 4 m × 4 m square plate completely clamped at four edges is divided into 40 × 40 fine quadrangular shell elements. A 0.8 m × 0.8 m square area at the center of the plate is supposed to be non-designable.

The material properties are defined as: elastic modulus $7 \times 10^{10}$ Pa, density 2,700 kg/m$^3$ and Poisson's ratio 0.3

The SIMP model with the penalty factor $p = 3$ is used here and a volume fraction of solid material is limited to 50%. In this case, the localized modes start to appear at the eighth iteration where the fundamental frequency decreases sharply. Later, some unsupported materials appear near the right corners on the top and bottom at the 16th and 17th iterations, as shown in Figure 2.8.

In Figure 2.9, the first mode shape is recognized as the localized mode at the 16th iteration and appears at the two corners. As the localized mode dominates the fundamental frequency, this will lead to meaningless iterations.

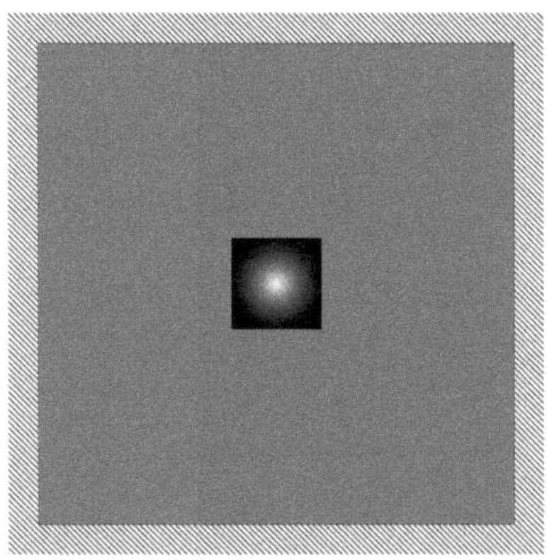

**Figure 2.7.** *Design domain of a square plate*

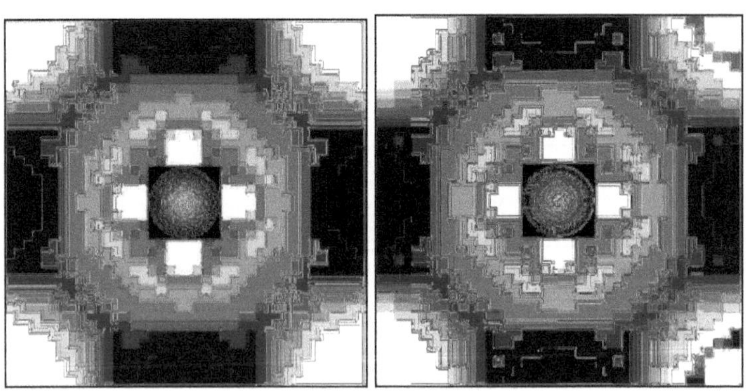

**Figure 2.8.** *The material layout at the: a) 16th and ; b) 17th iterations*

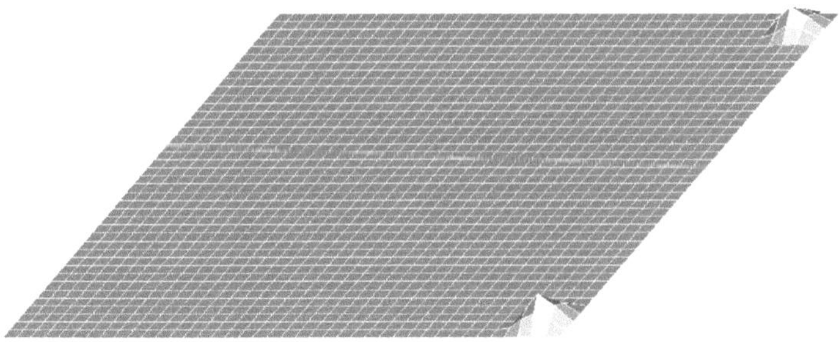

**Figure 2.9.** *The first mode shape at the 16th iteration*

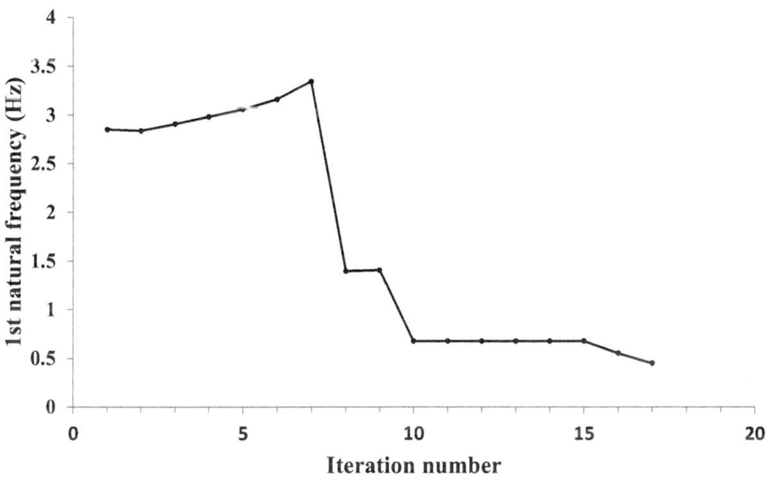


**Figure 2.10.** *The iteration history of the fundamental frequency*

Figure 2.10 shows the variation of the fundamental frequency during the iteration. A sudden decrease in the frequency value occurs at the eighth iteration because the localized mode is very different from the normal mode that occurs. Physically, the low-density areas are so compliant that a rather low fundamental frequency is obtained. Therefore, the phenomenon of localized mode is an ill-conditioned issue due to the mismatch between the mass and stiffness.

## 2.2. Localized deformation

Similar to the localized mode in vibration problems, the localized deformation often occurs in the presence of design-dependent loads. Here, design-dependent loads refer to those loads that change with respect to design variables, e.g. body force, inertia load and pressure load on a movable surface. Due to their existences, low-density areas are too weak to support themselves. Consequently, deformations in the low-density areas become much more important than those in the solid parts. Further discussions can also be found in the work of Bruyneel and Duysinx [BRU 04].

Here, new formulations are derived and numerical tests are made to highlight the problem. The same beam studied in section 2.1 is considered again. Suppose the beam undergoes a vertical gravity acceleration of 10 m/s$^2$. Material properties of the solid element are:

> Elastic modulus $10^{11}$ Pa, density 1,000 kg/m$^3$ and Poisson's ratio 0.3

Figure 2.11 shows that the gravity acceleration force is discretized into nodal loads over the beam. Elastic modulus of element 2 is interpolated with the SIMP model (penalty factor $p = 3$), while element 1 is a solid element that is supposed to be unchanged. The gravity acceleration force of element 2 is linearly penalized in terms of $\eta_2$ so that elastic modulus decreases more rapidly. Figure 2.12 plots the vertical displacements of nodes A and B versus $\eta_2$.

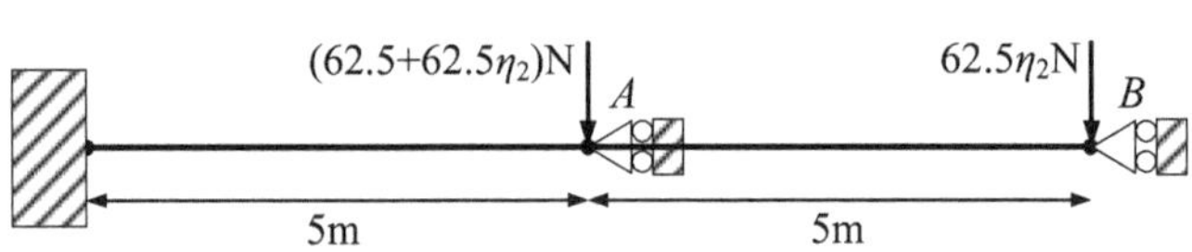


**Figure 2.11.** *The equivalent nodal loads when gravity is applied*

Both nodal loads vary linearly with $\eta_2$. As the material properties of element 1 remain unchanged with $\eta_1 = 1$, a linear relation exists between the displacement of node $A$ and $\eta_2$. Comparatively, as the stiffness of element 2 varies with the cubic exponent of $\eta_2$, the absolute displacement of node $B$ increases sharply as a relatively large localized deformation around $\eta_2 = 0.05$ and should be avoided.

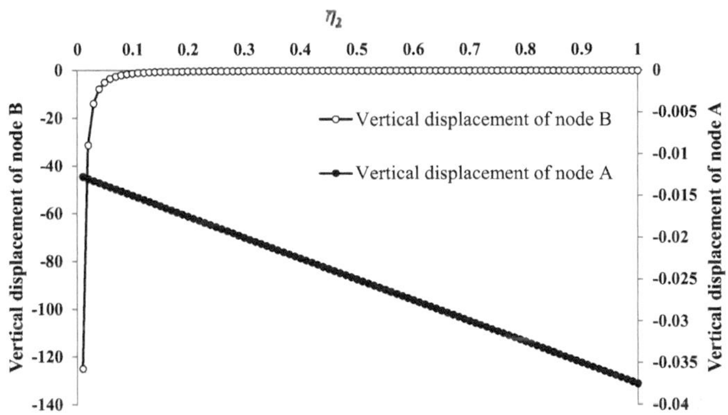


**Figure 2.12.** *Nodal displacements versus $\eta_2$*

Topology optimization of a two-dimensional (2D) problem with inertial load is considered. The standard SIMP model is used here with penalty factor $p = 3$. The design domain consists of $50 \times 100$ quadrangular finite elements, as shown in Figure 2.13. Because of the symmetry, only half of the domain is taken into account. Material properties are:

> Elastic modulus $7 \times 10^{10}$ Pa, density 2,700 kg/m$^3$ and Poisson's ratio 0.3.

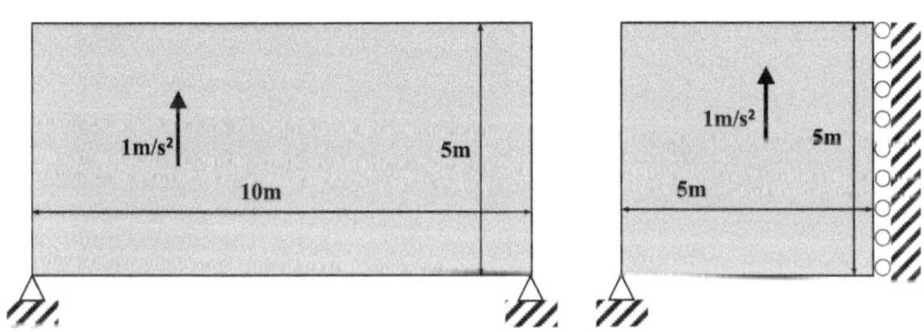


**Figure 2.13.** *Design domain of the test problem*

Figure 2.14(a) shows the material layout at the 13th iteration. Because the low-density areas are too compliant to support themselves, localized deformations appear at the bottom center, as shown in Figure 2.14(b). This implies that the localized elements become dominant and generate unconnected patterns in the following iterations. Figure 2.14(c) shows the singular pattern at the 19th iteration.

a)

b)

c)

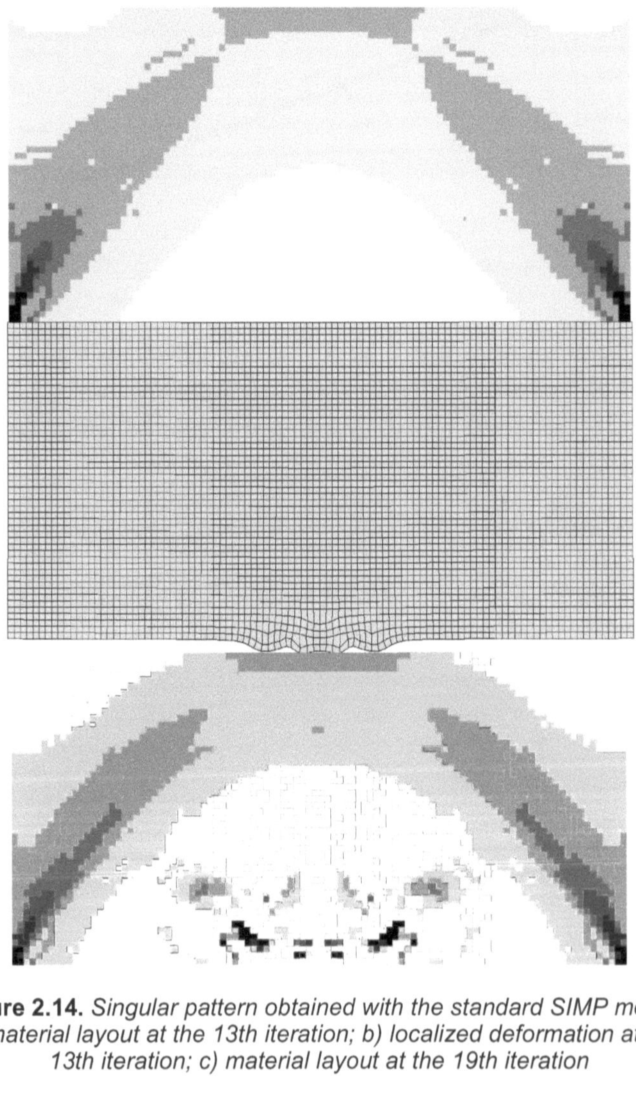

**Figure 2.14.** *Singular pattern obtained with the standard SIMP model: a) material layout at the 13th iteration; b) localized deformation at the 13th iteration; c) material layout at the 19th iteration*

## 2.3. Polynomial interpolation model

The SIMP model corresponds to a power-law interpolation expressed as:

$$\begin{aligned} \mathbf{K}_i &= P(\eta_i)\mathbf{K}_{i0} = \eta_i^p \mathbf{K}_{i0} \\ \mathbf{M}_i &= \eta_i \mathbf{M}_{i0} \end{aligned} \quad [2.15]$$

where the stiffness and the mass matrices depend upon pseudo-density variables in the exponential and linear form, respectively.

As discussed in sections 2.1 and 2.2, the SIMP model is not suitable for vibration problems and problems with design-dependent loads due to the mismatch penalty between the mass and stiffness. According to Pedersen [PED 00], a ratio of the pseudo-density variable to the penalty function is introduced as:

$$R_{MK} = \frac{\eta_i}{P(\eta_i)} \quad [2.16]$$

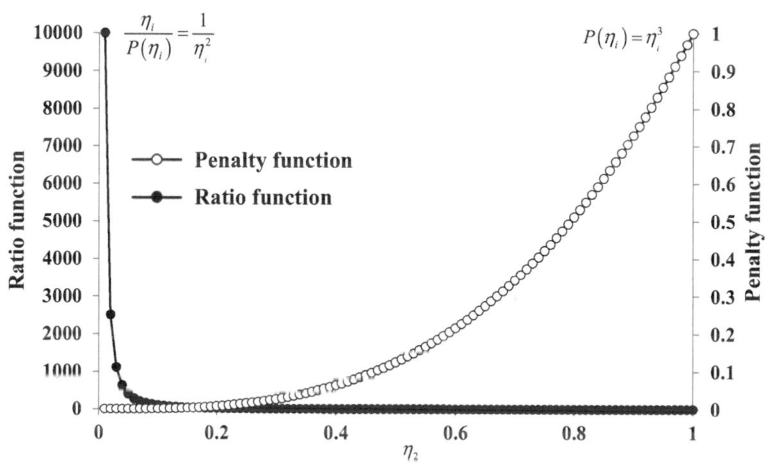


**Figure 2.15.** *Interpolation model and ratio function of standard SIMP*

Both the SIMP model ($p$ = 3) and the ratio function $R_{MK}$ are plotted with respect to the pseudo-density variable in Figure 2.15. The value of the ratio function tends to infinity when the pseudo-density variable approaches zero, which indicates the mismatch between the stiffness and mass matrices.

To avoid this problem, Pedersen [PED 00] constrained the lower bound of the elastic modulus to be 1/1,000 of the solid one.

$$\begin{aligned} P(1) &= 1 \\ P(0) &= 0.001 \end{aligned} \quad [2.17]$$

A cubic function is built accordingly as:

$$P(\eta_i) = \eta_i^3 + 0.001\left(1 - \eta_i^3\right) \quad [2.18]$$

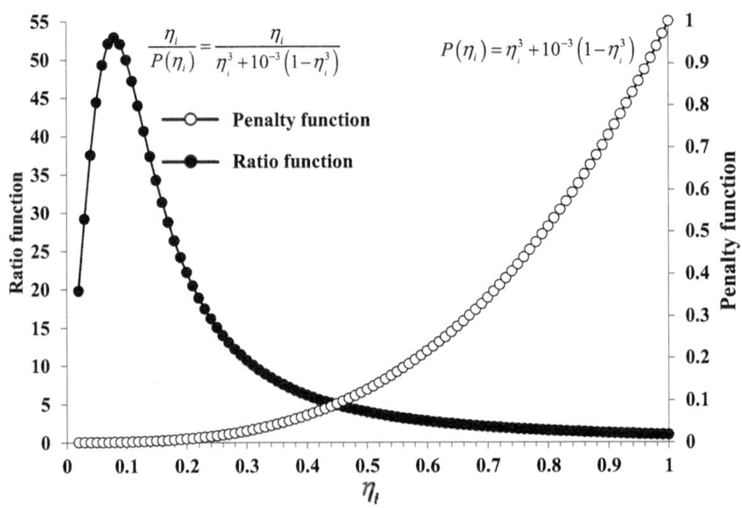


**Figure 2.16.** *Improved model and ratio function*

The modified function and the ratio function $R_{MK}$ are plotted in Figure 2.16. The value of the ratio function will no longer tend to the infinity when the pseudo-density variable approaches zero.

To a certain extent, this interpolation model can be used to solve eigenvalue maximization problems and avoid the localized modes.

However, the peak value of the ratio function around $\eta_i = 0.1$ is still unsatisfied and could possibly lead to localized modes.

Later, it was found that the microstructures of low densities can retain more stiffness than the SIMP model. Take a square shape microstructure "□" as an example. The equivalent tensile stiffness is calculated by virtue of the strain energy based method [PED 98, ZHA 07]. This method proved to be equivalent to the homogenization method in predicting the effective elastic material properties [HOR 99]. Let us take an example. The boundary conditions are defined in Figure 2.17. Horizontal displacements on the left edge are constrained. A uniform horizontal displacement $u$ is imposed on the right edge. The volume fraction of the microstructure is calculated as:

$$\eta = \frac{l_1^2 - l_2^2}{l_1^2} \qquad [2.19]$$

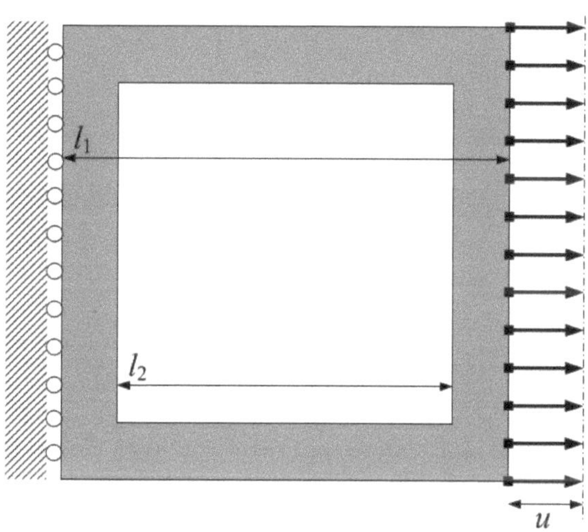


**Figure 2.17.** *Definition of the microstructure and the boundary conditions*

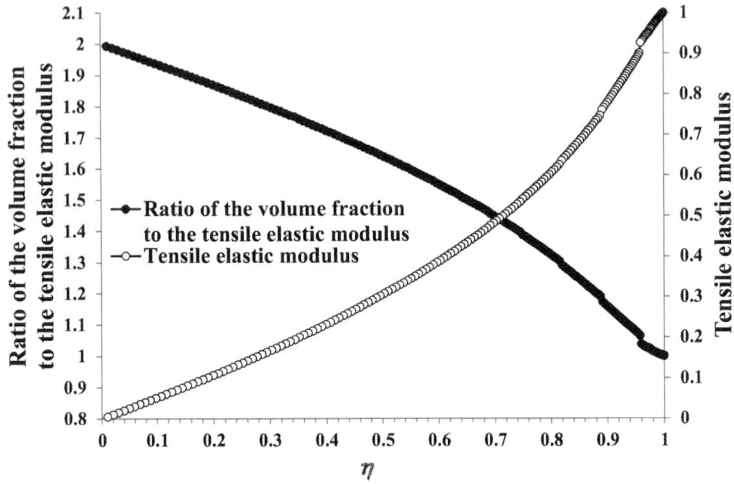


**Figure 2.18.** *The horizontal tensile elastic modulus versus the volume fraction of the "□" shaped microstructure*

The tensile elastic modulus is calculated as:

$$\frac{Eu}{l_1} = \frac{f_R}{l_1 t} \qquad [2.20]$$

$$E = \frac{f_R}{ut} \qquad [2.21]$$

where $E$ is the equivalent tensile elastic modulus, $f_R$ is the resultant of reaction forces on the left edge. $t$ is the thickness of the microstructure. Assume that $l_1 = u = t = 1$ and $l_2$ varies from 0.005 to 0.995 in the test. The relationships between the tensile elastic modulus and the volume fraction are plotted in Figure 2.18. Here, the ratio just corresponds to the ratio function $R_{MK}$.

In comparison to the curves related to the SIMP in Figure 2.15, the slope of the tensile elastic modulus is positive at $\eta = 0$ and the ratio of the volume fraction to the tensile elastic modulus is limited to a small value approximately equal to 2.

The maximum value of the ratio can be proved by ignoring the two vertical columns of the microstructure when $\eta$ is rather small. As a result, the resultant reaction forces are approximately calculated as the inner tensile forces of the two horizontal bars.

$$\frac{f_R}{(l_1 - l_2)t} = \frac{E_0 u}{l_1} \quad [2.22]$$

where $E_0$ is the elastic modulus of the solid material forming the microstructure. The equivalent tensile elastic modulus is then calculated by

$$\frac{Eu}{l_1} = \frac{f_R}{l_1 t} = \frac{E_0 u(l_1 - l_2)}{l_1^2} \quad [2.23]$$

$$E = \frac{E_0(l_1 - l_2)}{l_1} \quad [2.24]$$

And the ratio of the volume fraction to the tensile elastic modulus is expressed as:

$$\lim_{\eta \to 0} \frac{\eta}{E} = \lim_{l_2 \to l_1} \frac{(l_1^2 - l_2^2)}{l_1^2} \cdot \frac{l_1}{E_0(l_1 - l_2)} = \lim_{l_2 \to l_1} \frac{l_1 + l_2}{l_1 E_0} = \frac{2}{E_0} \quad [2.25]$$

In this example, $E_0$ is predefined as 1. At $\eta = 0$, the derivative of $E$ with respect to $\eta$ is 0.5, which makes the curve different from that of the SIMP model. However, the homogenization method involves greatly computational complexities.

RAMP [STO 01] is an alternative interpolation model. The penalty function and its derivative at $\eta_i = 0$ are expressed as:

$$P(\eta_i) = \frac{\eta_i}{1 + q(1 - \eta_i)} \quad [2.26]$$

$$\lim_{\eta_i \to 0} P'(\eta_i) = \lim_{\eta_i \to 0} \frac{1+q}{(1+q-q\eta_i)^2} = \frac{1}{1+q} \qquad [2.27]$$

where $q$ is the penalty factor, which is always positive. Figure 2.19 shows the penalty function and the ratio function at $q = 5$. It is obvious that $R_{MK}$ is actually a linear function of pseudo-density variable.

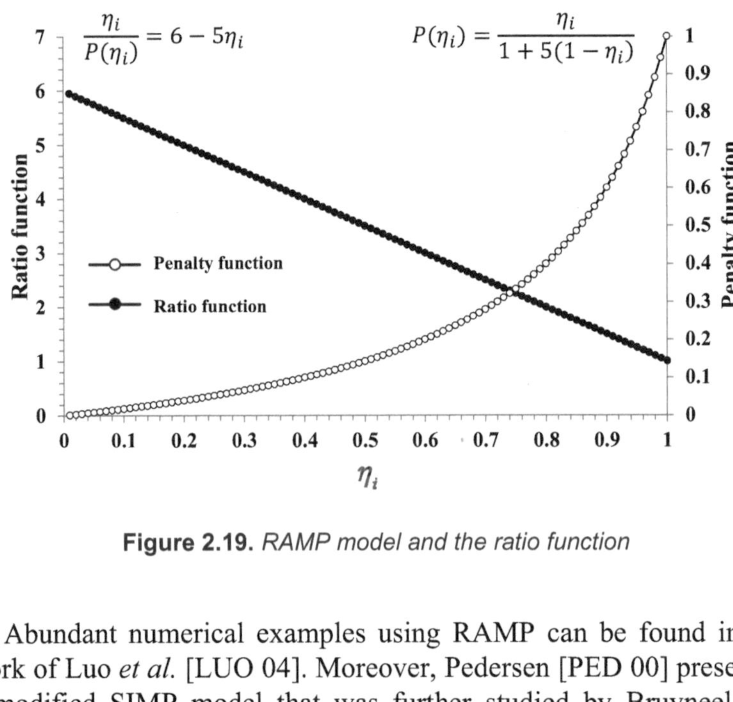


**Figure 2.19.** *RAMP model and the ratio function*

Abundant numerical examples using RAMP can be found in the work of Luo *et al.* [LUO 04]. Moreover, Pedersen [PED 00] presented a modified SIMP model that was further studied by Bruyneel and Duysinx [BRU 04].

$$E_i = P(\eta_i)E_{i0} = \begin{cases} (\eta_i/\alpha)E_{i0}; & \eta_i \le \alpha^{\frac{1}{1-p}} \\ \eta_i^p E_{i0} & ; \eta_i > \alpha^{\frac{1}{1-p}} \end{cases} \qquad [2.28]$$

where two parts exist for the penalty function. They are linear interpolation with a slope of $1/\alpha$ and power-law interpolation of the standard SIMP model. The linear penalty is defined to ensure the positive derivative near $\eta_i = 0$. Two independent parameters $\alpha$ and $p$ can adjust the $R_{MK}$ and the penalty effect separately.

In Figure 2.20, both the penalty function and the ratio function are plotted at $\alpha = 16$ and $p = 3$. Thus, the maximum value of $R_{MK}$ is limited by the constant value of $\alpha$ to control the mismatch between the mass and stiffness.

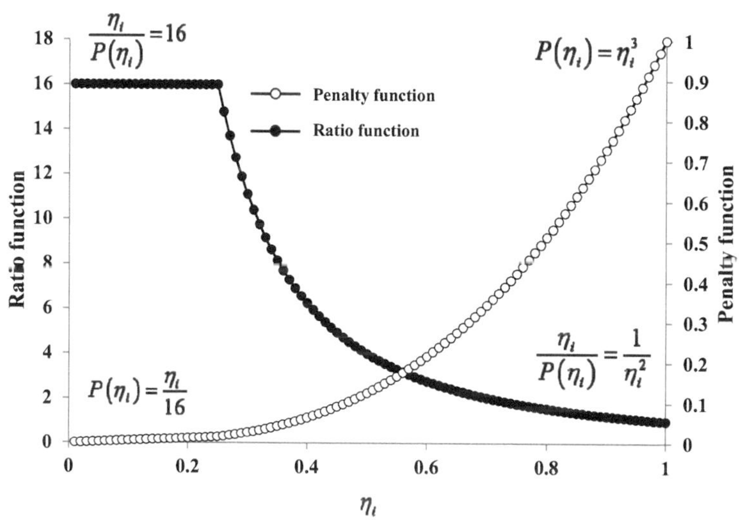


**Figure 2.20.** *Modified SIMP model and the ratio function*

The modified SIMP model was found to be effective for topology optimization of problems with body force or natural frequency maximization. However, critical comments are received for the non-differentiability at $\eta_i = \alpha^{1/(1-p)}$. Here, a polynomial interpolation function [ZHU 09] is presented in the following form.

$$P(\eta_i) = (1-w)\eta_i^p + w\eta_i \qquad [2.29]$$

This function holds the following properties:

1) at two end points

$$P(0) = 0,\ P(1) = 1$$

2) the first-order derivative at $\eta_i = 0$

$$P'(0) = w, \ w > 0$$

Figure 2.21 shows the new penalty function and the ratio function, respectively. $R_{MK}$ is limited to the value $1/w$ at $\eta_i = 0$ and the derivative of the penalty function is continuous everywhere to favor gradient-based optimization methods.

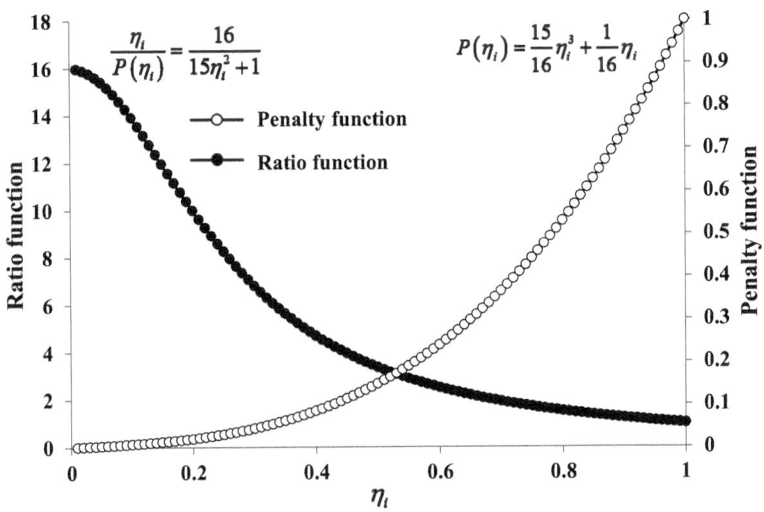


**Figure 2.21.** *Polynomial interpolation model and the ratio function*

Now, the polynomial interpolation model is examined based on the previously tested two examples. For the vibrating square plate shown in Figure 2.7, material layouts and the mode shape at the 30th iteration are shown in Figure 2.22(a). It is obvious that the mode shape is a global mode without localized mode.

The optimization is further processed and convergence attained at the 42nd iteration. The non-designable area is finally supported as shown in Figure 2.22(b). Figure 2.23 indicates that the convergence history of the objective function is rather stable. A similar optimized layout can be found in Pedersen [PED 00].

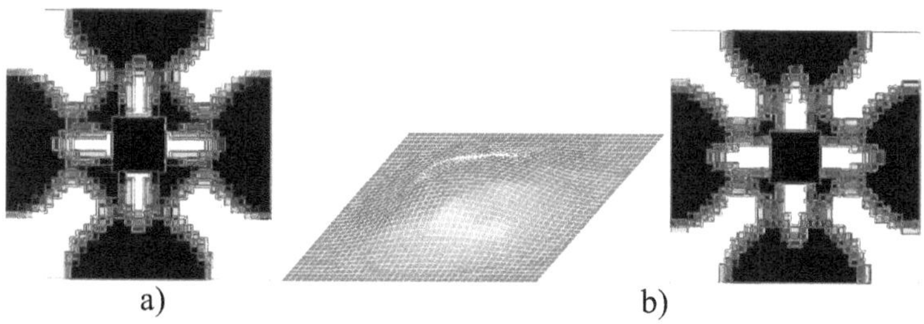


**Figure 2.22.** *Layout design of the square plate with polynomial interpolation model: a) material layout and mode shape at the 30th iteration; b) material layout at thc 42nd iteration*

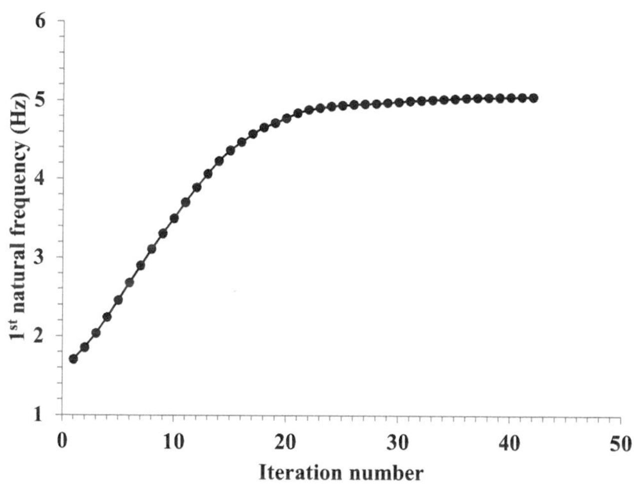


**Figure 2.23.** *Convergence history of the first natural frequency*

Again, the problem with inertial load discussed in Figure 2.13 is tested here. One intermediate solution and the corresponding dcformation shapc arc shown in Figurcs 2.24(a). There is no localized deformation found in the design domain. By continuing the design iteration, the optimized layout is obtained at the 38th iteration, which is a typical arc illustrated in Figures 2.24(b). Figure 2.25 shows that the convergence history of the global strain energy is also rather stable. Similar topology optimization problems can be found in [BRU 05].

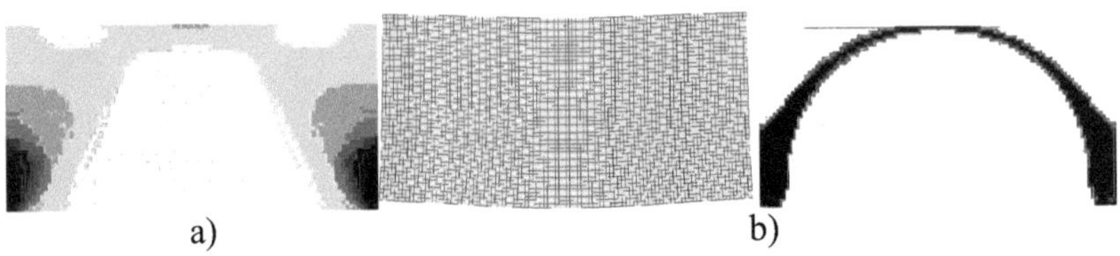


**Figure 2.24.** *Layout design of the inertia load problem with polynomial interpolation model: a) material layout and global deformation at the 16th iteration; b) material layout at 38th iteration*

According to numerical results, the polynomial interpolation model is effective in avoiding localized modes and localized deformations. This will benefit topology optimization with vibrations and design-dependent body forces, and will be further highlighted in the following chapters.

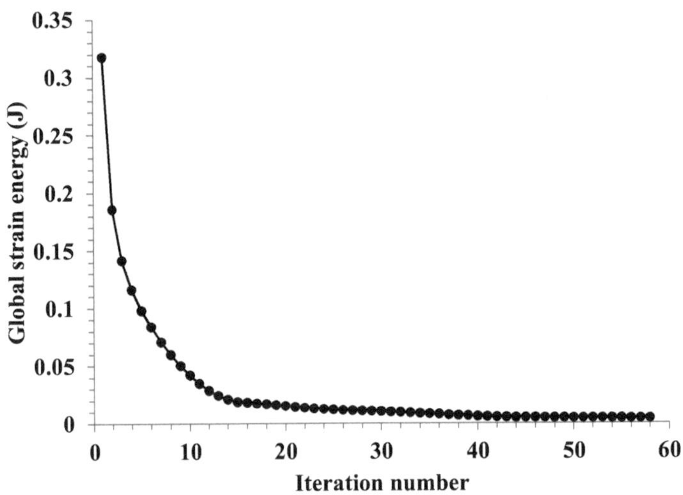


**Figure 2.25.** *Convergence history of the global strain energy*

## 2.4. Breakdown issue in ESO

The topic discussed in this section is an important issue involved in the family of evolutionary structural optimization (ESO) methods from the viewpoint of low-density areas. Its implementation is easily understood with less mathematical rigor. Based on the heuristic engineering concepts and sensitivity results, the material layout of a structure is optimized by removing inefficient elements systematically from the ground structure. Numerical results have shown that a variety of problems, e.g., maximizations of structural stiffness [XIE 94a], natural frequency [XIE 94b, XIE 96] and buckling load [RON 01] can be dealt with by this method. However, the breakdown issue discovered by Zhou and Rozvany [ZHO 01] is still a fatal drawback of this method.

ESO/BESO method works directly with 0–1 discrete design variables. Element deletions and recoveries are carried out by virtue of the element efficiency which is measured by the sensitivity values. As we cannot directly obtain all the sensitivity values of the finite difference with only one finite element analysis, sensitivity values used in the hard-killing scheme of ESO/BESO are analytical sensitivities.

$$\frac{\partial C}{\partial \eta_i} = -\frac{1}{2}\mathbf{U}_i^{\mathrm{T}}\mathbf{K}_i\mathbf{U}_i = -C_i \qquad [2.30]$$

$$\frac{\partial \omega^2}{\partial \eta_i} = \frac{\mathbf{U}_i^{\mathrm{T}}\mathbf{K}_i\mathbf{U}_i - \omega^2\mathbf{U}_i^{\mathrm{T}}\mathbf{M}_i\mathbf{U}_i}{\mathbf{U}^{\mathrm{T}}\mathbf{M}\mathbf{U}} \qquad [2.31]$$

where $C_i$ is the strain energy of the $i$th element. The sensitivities can be derived by supposing a linear interpolation model for $\mathbf{K}_i$ and $\mathbf{M}_i$ as a function of $\eta_i$ with

$$\begin{aligned} \mathbf{K}_i &= \eta_i\mathbf{K}_{i0} \\ \mathbf{M}_i &= \eta_i\mathbf{M}_{i0} \end{aligned} \qquad [2.32]$$

$\mathbf{K}_{i0}$ and $\mathbf{M}_{i0}$ denote the stiffness and mass matrices of the $i$th element when it is solid. In fact, the analytical sensitivities are unable to represent the real change of the objective function whenever an inefficient element is directly removed ($\Delta\eta_i = -1$) or an efficient element is recovered ($\Delta\eta_i = 1$). To solve this problem, consider two kinds of sensitivities shown in Figure 2.26.

The solid curve in Figure 2.26 represents the objective function versus $\eta_i$. This is typically the case for the structural strain energy with values of other design variables being invariable. $\Delta C_1$ corresponds to the change of $C$ evaluated with the analytical sensitivity value when the $i$th element is removed by ESO/BESO from the design domain, whereas $\Delta C_2$ represents the exact change of $C$ after the removal of the $i$th element. Clearly, both are quite different from each other.

The change of the objective function will be overestimated when $\Delta C_1 > \Delta C_2$. This is a conservative situation so that some inefficient elements are retained in the actual iteration. In contrast, when $\Delta C_1 < \Delta C_2$, particularly when a sharp difference exists, there is the danger that related elements could be erroneously considered to be inefficient, removed and no longer recoverable during the iterations.

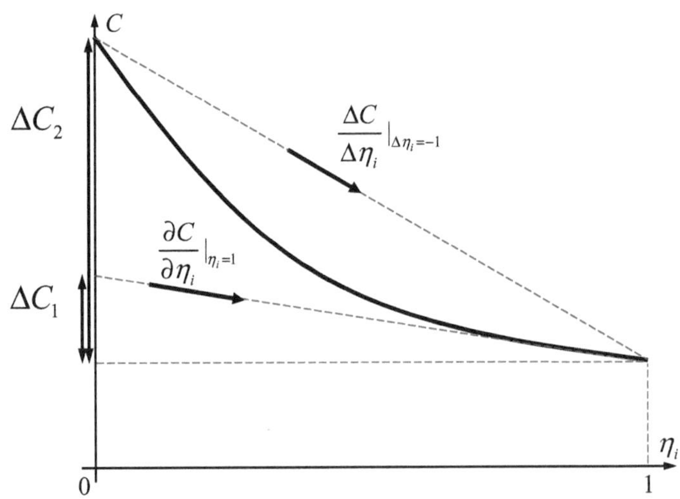


**Figure 2.26.** *Comparison of two kinds of sensitivities*

**Figure 2.27.** *A test example for the failure of ESO*

Although lots of existing results have shown that the convergence of the iteration procedure is not deteriorated by ESO/BESO, understanding the nature of the solid curve in Figure 2.26 is essential to reveal the underlying trouble of ESO/BESO.

To illustrate the phenomenon and reasons for the failure of ESO, a test example from Zhou and Rozvany [ZHO 01] is analyzed here in detail. The FE model shown in Figure 2.27 consists of a 32 m × 3 m horizontal beam and a 1 m × 4 m vertical link meshed with 0.25 m × 0.25 m 4-node quadrangular elements. The material properties are:

Elastic modulus 1 Pa, Poisson's ratio 0

To figure out the difficulty, the initial distribution of elemental strain energies representing the absolute sensitivity values is shown in Figure 2.28(a). It is found that the elements on the vertical link take the lowest strain energy and should be removed in the ESO procedure. Based on the reanalysis of the updated model, Figure 2.28(b) shows a new distribution of the element strain energies with a sudden increase of the structural strain energy $C$ by a factor of 10. This implies that a fatal iteration occurs. As the elements with the maximum strain energies are unfortunately relocated at the left bottom corner of the horizontal part, erroneously removed elements on the vertical link are no longer recovered by BESO.

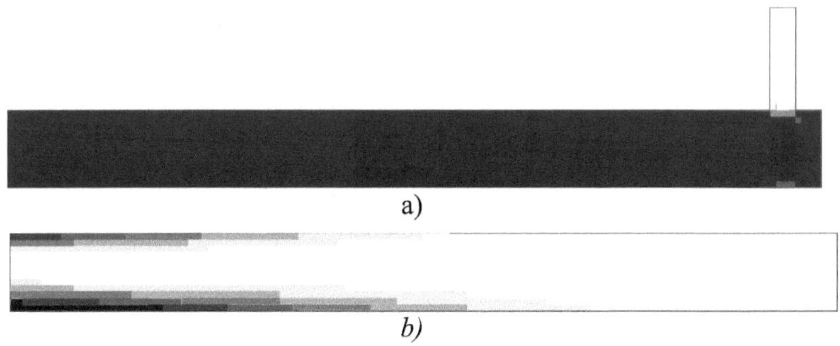


**Figure 2.28.** *Distributions of the elemental strain energies related to ESO method: a) initial distribution of elemental strain energies within the full structure (C = 194.9 J); b) distribution of elemental strain energies after removing the vertical link (C = 2306.1 J)*

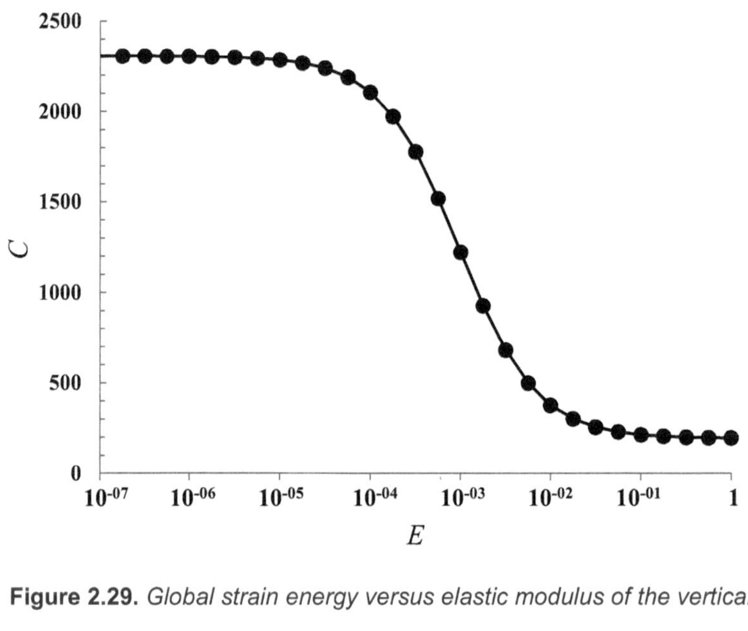


**Figure 2.29.** *Global strain energy versus elastic modulus of the vertical link*

To obtain the sudden variation of the objective function, the strain energy of the total structure $C$ is drafted versus the variation of the elastic modulus of elements $E$ from 1 to $10^{-7}$ Pa on the vertical link in Figure 2.29. Here, the axis of $E$ is logarithmic. It is observed that the slope is very small near $E = 1$ and $E = 0$. This indicates that both solid and void elements are not sensitive. As a result, the solid elements are removed in ESO and cannot be recovered by BESO.

However, a significant change of $C$ is observed for values of $E$ between $10^{-2}$ and $10^{-5}$ Pa. However, the sensitivity values used in ESO cannot detect the sharp variation of the global strain energy as discussed before. Thus, the vertical link is apparently considered to be inefficient and completely removed. This is the reason why the ESO fails.

According to the above discussions, the standard ESO/BESO approach cannot detect the sharp increase in the strain energy of the vertical link. Figure 2.29 illustrates high absolute sensitivity values when $E$ varies between $10^{-2}$ and $10^{-5}$ Pa. This means that the elements on the vertical link should take the maximum strain energies of all the elements for a wide range of $E$. To make things clear, Figure 2.30 shows the variation of the strain energy of the vertical link in terms of its elastic modulus, varying from 1 to $10^{-7}$ Pa.

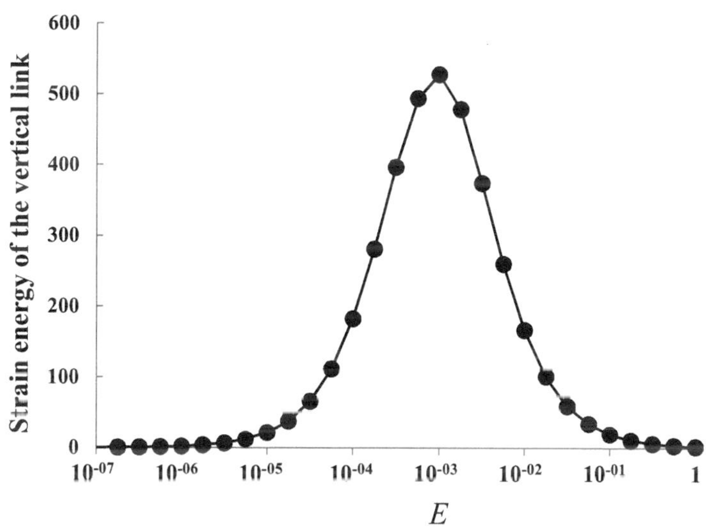


**Figure 2.30.** *Strain energy of the vertical link versus its elastic modulus*

Accordingly, the so-called check position method is proposed to identify and remedy the erroneous element deletion. The key idea is to introduce moderate low-density elements with the elastic modulus set to 0.01–0.1% of the initial value, i.e. $10^{-4}$–$10^{-3}$ Pa in this problem. These elements are able to capture the sharp variation of the objective function. Compared to solid elements, moderate low-density elements will act as compromising elements between void and solid elements but they are not compliant enough to be treated as removed ones. This makes it possible to pick out the elements that were erroneously removed and to recover them in the design procedure.

In the test example, elements on the vertical link are now replaced with the moderate low-density elements of equivalent elastic modulus $10^{-3}$ Pa. The corresponding distribution of the elemental strain energies is shown in Figure 2.31. It is observed that vertical link substituted with the moderate low-density element takes the maximum value of the elemental strain energies, which is easily identified in the design domain.

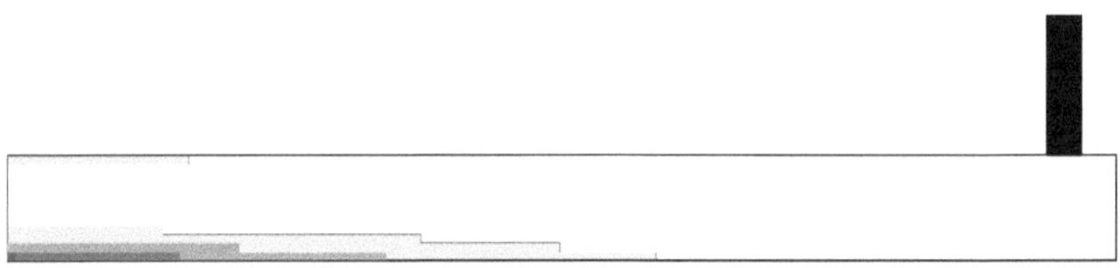

**Figure 2.31.** *Distribution of the elemental strain energies at the check position*

In this way, a modified ESO/BESO design procedure can be devised as:

1) calculate the design sensitivities and remove the inefficient elements;

2) if a sharp degradation of the objective function is detected against a specified threshold, e.g. the structural strain energy increases to more than 5 times or the natural frequency decreases to less than 1/5, the removed elements in the last step will be further recovered with the moderate low-density elements. This is referred to as the

check position. Otherwise, a normal ESO/BESO procedure of element removal and growth is carried out;

3) based on the FE reanalysis of the updated model at the check position, moderate low-density elements possessing the maximum strain energies will be recovered and marked as non-designable, while the other moderate elements will be removed again. But the mark of non-designable elements only stays for a few steps. These elements will be designable again later on;

4) standard ESO/BESO element removal and growth procedure is carried out by ignoring the non-designable elements.

The test example is now continued with this procedure. With the vertical link recovered and marked as non-designable, the elements removed in the first effective step are presented in Figure 2.32.

However, after several iterations, when the elements on the vertical link are once again designable, the elements can be removed again. The check position may be reintroduced depending on the objective degradation. After 61 iterations, the final design of the structure is shown in Figure 2.33 with an amount of 40% of the total material.

**Figure 2.32.** *Structural topology generated by the first effective iteration*

**Figure 2.33.** *Optimal structural topology (C = 509.8 J)*

Another test is to maximize the first natural frequency. As shown in Figure 2.34, a 2 m × 12 m beam supported at both ends is meshed with 20 × 120 quadratic elements. The non-designable area consists of 2 × 20 elements at the center. Fifty percent of the total material is used. Material properties are

elastic modulus $E_0 = 2\times10^{11}$ Pa, density $\rho_0 = 7,800$ kg/m$^3$
and Poisson's ratio $\mu = 0.3$

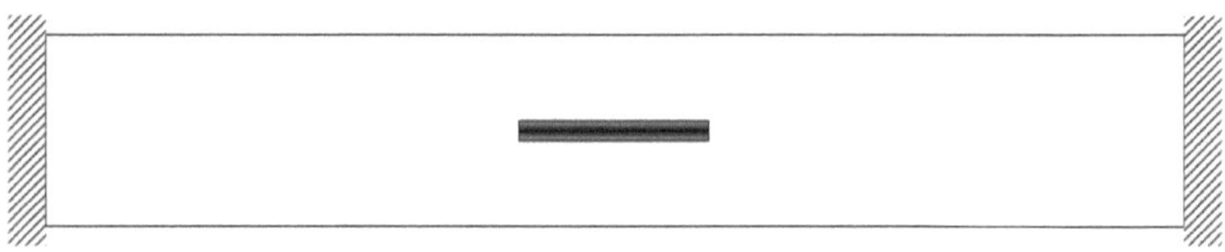

**Figure 2.34.** *Design domain of the biclamped beam*

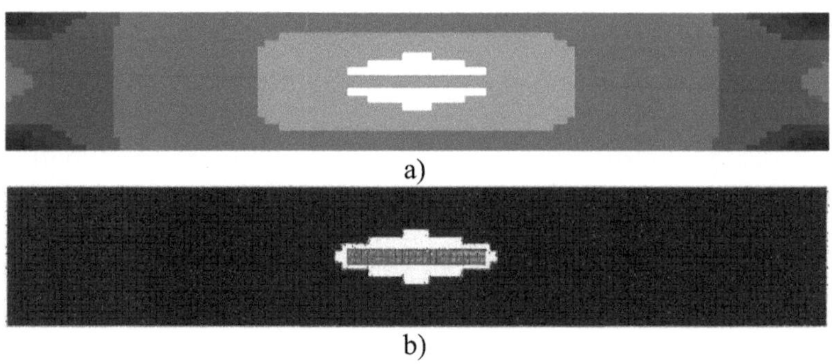

**Figure 2.35.** *Erroneous iterations: a) distribution of the sensitivities at the third iteration (64.17 Hz); b) structural topology at the fourth iteration (0 Hz)*

Problems are detected at the fourth iteration where the non-designable area is unsupported and the objective function decreases to 0 Hz, as shown in Figure 2.35.

The removed elements in the last iteration are then recovered with the moderate low-density elements. The updated structure is analyzed again and the distribution of sensitivity values is shown in Figure 2.36. It is observed that some of the moderate low-density elements take now the maximum values and will be recovered into the solid elements, as shown in Figure 2.37.

**Figure 2.36.** *Distribution of sensitivity values at the check position (59.19 Hz)*

Sixty-eight iterations are completed and the optimized design is shown in Figure 2.38. The proposed ESO/BESO design procedure with check position can be used to avoid some erroneous design iterations. However, this method is still based on some heuristic rules.

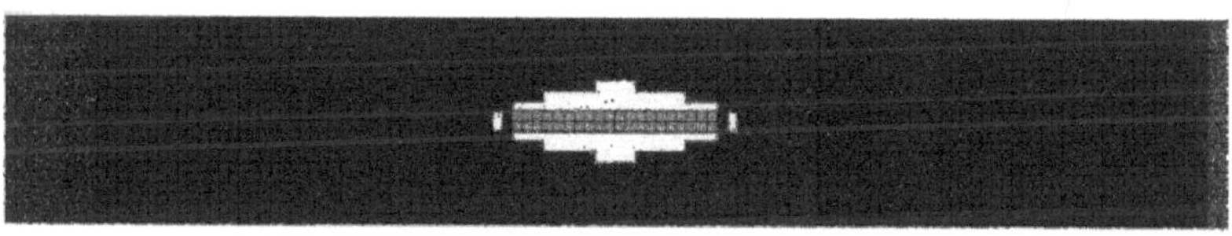

**Figure 2.37.** *The effective structural topology at the fourth iteration (64.27 Hz)*

**Figure 2.38.** *The final design of the biclamped beam (71.18 Hz)*

## 2.5. Conclusions

To benefit from the applications of topology optimization, some problems involved in the low-density areas in topology optimization are discussed in detail. By observing the phenomena of the localized modes in eigenvalue maximization problems and localized deformations in problems with design-dependent body forces, numerical failures are concluded to be the result of improper interpolation models. The great difference between mass and stiffness of the low-density elements will thus be limited to avoid these problems. Several interpolation models, e.g. limiting the minimum value of the elastic modulus or the derivative of the interpolation, are evaluated. A polynomial interpolation model is also presented and discussed through numerical tests.

Finally, the breakdown of the ESO/BESO method is discussed. It is found that the sensitivity values cannot correctly describe the practical operations on the elements. The check position scheme is then proposed by introducing moderate low-density elements into design iterations. Consequently, erroneously removed elements can be identified and recovered. Two typical examples are examined to illustrate how to achieve reasonable results.

# 3

# Dynamic Problems

## 3.1. Introduction

Topology optimization of dynamic problems is a challenging topic, and mainly includes optimization problems related to structural eigenfrequency and dynamic responses.

Relevant research on structural eigenfrequency optimization have been focused on free vibration problems, such as the maximization of fundamental frequency and the gap between two consecutive natural frequencies. For example, based on the homogenization method, Díaz and Kikuchi [DÍA 92] and Ma *et al.* [MA 95] studied the maximization of the fundamental frequency and a set of eigenvalues of the structure, respectively. Later, Pedersen [PED 00] studied the maximization of eigenvalues using the density method. In his work, specific attention was paid to the localized modes in low-density areas and an improved interpolation scheme based on SIMP was developed to solve this problem. Also using the density method, Du and Olhoff [DU 07] dealt with the topology optimization involving the maximization of eigenfrequencies and frequency gaps. In their work, another improved interpolation scheme based on [TCH 02] was used to avoid the localized modes. The RAMP model introduced by Stolpe and Svanberg [SVA 01] and the polynomial interpolation model introduced by Zhu *et al.* [ZHU 09] were also implemented to eliminate the localized modes. These research achievements concerning localized modes have benefited the study of dynamic response optimization.

Dynamic response optimization mainly includes problems related to harmonic responses [JOG 02, KAN 12, LIU 15, MA 95, OLH 05, YAN 14] and random vibration responses [LIN 11, RON 00, ZHA 10a, ZHA 12, ZHA 15]. Physically, the minimization of the dynamic response subjected to harmonic excitations is a basic concern since the major source of vibration or noise in a structure system or a machine approximately corresponds to a harmonic force due to rotating components. For example, practical implications range from household appliances and construction machinery to cars and ships. All kinds of periodic excitations can be interpreted as the superposition of a set of harmonic excitations.

Until now, considerable effort has been made to study topology optimization for harmonic responses. Ma *et al.* [MA 95] defined the "dynamic compliance" as the objective function without considering structural damping, and successfully implemented the homogenization technique for topology optimization under harmonic excitations. Jog [JOG 02] also minimized the "dynamic compliance" for which structural damping was taken into account. The frequency response amplitude at a given point was also optimized in his work. Tcherniak [TCH 02] designed the layout of resonating actuators by maximizing the magnitude of steady-state vibrations at a given excitation frequency. Yoon [YOO 10] investigated the applicability of model reduction techniques in topology optimization of harmonic problems, including the mode displacement method (MDM), the Ritz vector method and the quasi-static Ritz vector method. Shu *et al.* [SHU 11] minimized the frequency response based on the level set method. Kang *et al.* [KAN 12] studied the optimal distribution of damping material in vibrating structures subjected to multiple harmonic excitations in phase, using the topology optimization method.

Random excitations such as earthquake ground motions, ocean wave-induced forces, aerodynamic and turbulent pressures can be classified into stationary and non-stationary random excitations. In general, if the overall averages for a random excitation were time-independent, the excitation would be stationary. Otherwise, it would be non-stationary. To a large extent, some secondary structure designs are mostly based on stationary random excitations [WIJ 09] that were closely considered in the advanced formulations of dynamic optimization problems [BUC 98, MA 11, MIS 13, PAG 12].

Rong *et al.* [RON 00] optimized the structural topology using the ESO method with stationary random responses constrained in design. Dynamic responses were calculated by means of the complete quadratic combination method (CQC). Comparatively, as CQC was cost-ineffective in random analysis, the pseudo-excitation method (PEM) [JIA 92, LIN 01, LIN 85, LU 09, ZHA 10b] was thus introduced to transfer the solving of random responses into the solving of pseudo-harmonic responses. Although both methods can achieve the same solution with the same number of structural modes, the efficiency of the PEM is much higher than the CQC. Lin *et al.* [LIN 11] adopted the PEM as an efficient optimization procedure in the maximization of energy harvesting performance under stationary random excitation.

However, structures were mostly limited to a small number of DOFs due to the inherent complexity of the problem in the previous work. How to deal with dynamic topology optimizations of large-scale problems remains a great challenge, even though different topology optimization formulations and methods have been developed.

In this chapter, linear dynamic systems with classical damping and steady-state responses are considered. A comparative study of different dynamic analysis methods is first made to highlight their computing accuracy and efficiency for problems under harmonic force excitations [LIU 15]. Their effectiveness in topology optimization under harmonic force excitations at one specific frequency and with multiple frequencies is, respectively, investigated, especially for large-scale problems. Investigations are also made into structural topology optimization related to dynamic responses under stationary random force excitations [ZHA 15]. It is shown that the commonly used CQC in previous optimization work is not only computationally costly, but also results in a non-convergent design pattern due to the low computing accuracy of random responses for large-scale problems. To circumvent these difficulties, an efficient and accurate optimization procedure integrating the PEM and mode acceleration method (MAM) is introduced into the dynamic topology optimization. In this framework, random responses are calculated using the PEM to ascertain a high efficiency over the CQC. More importantly, the accuracy of random responses is improved indirectly by solving the pseudo-harmonic responses involved in the PEM with the help of the MAM. Numerical examples fully demonstrate the validity of the developed optimization procedure and its potential applications in practical designs.

## 3.2. Analysis methods for harmonic force excitations

Although harmonic response analysis methods such as MDM, MAM [COR 83] and full method (FM) were widely implemented in commercial CAE software, how to use them correctly in topology optimization is still a basic issue to be clarified in practice. Generally, MDM is adopted by default due to its simplicity and efficiency from the engineering viewpoint. However, the poor response accuracy would affect the optimization convergence, especially for large-scale problems. Therefore, MAM and FM are introduced into the framework of topology optimization. MAM has a higher accuracy than MDM in the case of using the same modes and can easily be implemented. FM is an exact analysis method and can be used as a benchmark. A comparative study among MDM, MAM and FM is presented to highlight their effectiveness in topology optimization under harmonic force excitation for large-scale problems.

Here, linear dynamic systems with classical damping and steady-state responses under harmonic force excitations are considered. As is known, the governing equation of a discretized $n$-DOF structure under harmonic force excitation can generally be written as:

$$\mathbf{M}\ddot{\mathbf{u}}(t) + \mathbf{C}\dot{\mathbf{u}}(t) + \mathbf{K}\mathbf{u}(t) = \mathbf{f}(t) \quad [3.1]$$

where **M**, **C** and **K** represent the mass matrix, damping matrix and stiffness matrix. $\mathbf{u}(t)$ represents the displacement vector. $\mathbf{f}(t)$ denotes the harmonic force vector of form $\mathbf{f}(t) = \mathbf{F}e^{\imath\omega t}(\imath^2 = -1)$. **F** and $\omega$ denote the magnitude vector of harmonic force and excitation frequency, respectively. In this section, three typical methods, i.e. MDM, MAM and FM, are briefly introduced before studying their applications in topology optimization.

### 3.2.1. *Mode displacement method*

Suppose $\omega_k$ and $\boldsymbol{\varphi}_k$ are the $k$th circular eigenfrequency and eigenvector, respectively. The mode shape matrix $\boldsymbol{\varphi} = [\boldsymbol{\varphi}_1 \ldots \boldsymbol{\varphi}_n]$ is normalized by mass matrix. Classical damping is supposed, with $\zeta_k$ being the $k$th damping ratio so that following relations hold

$$\begin{cases} \boldsymbol{\varphi}^{\mathrm{T}}\mathbf{M}\boldsymbol{\varphi} = \mathbf{I} \\ \boldsymbol{\varphi}^{\mathrm{T}}\mathbf{K}\boldsymbol{\varphi} = diag(\omega_k^2) \\ \boldsymbol{\varphi}^{\mathrm{T}}\mathbf{C}\boldsymbol{\varphi} = diag(2\zeta_k\omega_k) \end{cases} \quad [3.2]$$

By introducing the notation,

$$\mathbf{u}(t) = \boldsymbol{\varphi}\mathbf{y}(t) \quad [3.3]$$

where $\mathbf{y}(t)$ is the vector of generalized coordinates. A number $n$ of uncoupled equations of motion can be obtained by substituting equation [3.3] into equation [3.1] and by premultiplying $\boldsymbol{\varphi}^{\mathrm{T}}$.

$$\ddot{y}_k(t) + 2\zeta_k\omega_k\dot{y}_k(t) + \omega_k^2 y_k(t) = \boldsymbol{\varphi}_k^{\mathrm{T}}\mathbf{f}(t) \quad [3.4]$$

The solution of the above equation yields

$$y_k(t) = (\omega_k^2 - \omega^2 + 2\imath\zeta_k\omega_k\omega)^{-1}\boldsymbol{\varphi}_k^{\mathrm{T}}\mathbf{f}(t) \quad [3.5]$$

Thus, the displacement response related to equation [3.3] under harmonic force can be expressed as:

$$\mathbf{u}(t)=\sum_{k=1}^{n}\boldsymbol{\varphi}_k y_k(t)=\sum_{k=1}^{n}\boldsymbol{\varphi}_k H_k \boldsymbol{\varphi}_k^{\mathrm{T}}\mathbf{f}(t) \qquad [3.6]$$

$H_k$ is the frequency domain transfer function between loading and response:

$$H_k=(\omega_k^2-\omega^2+2\iota\zeta_k\omega_k\omega)^{-1} \qquad [3.7]$$

Considering the computing efficiency, only lower $l$ modes are usually employed with $l<<n$ so that:

$$\mathbf{u}(t)=\sum_{k=1}^{l}\boldsymbol{\varphi}_k H_k \boldsymbol{\varphi}_k^{\mathrm{T}}\mathbf{f}(t) \qquad [3.8]$$

### 3.2.2. *Mode acceleration method*

The solution of the uncoupled system in equation [3.4] is rewritten as:

$$y_k(t)=\frac{\boldsymbol{\varphi}_k^{\mathrm{T}}\mathbf{f}(t)}{\omega_k^2}-\frac{2\zeta_k\dot{y}_k(t)}{\omega_k}-\frac{\ddot{y}_k(t)}{\omega_k^2} \qquad [3.9]$$

The substitution of equation [3.9] into equation [3.6] leads to:

$$\mathbf{u}(t)=\sum_{k=1}^{l}\left(\frac{\boldsymbol{\varphi}_k\boldsymbol{\varphi}_k^{\mathrm{T}}\mathbf{f}(t)}{\omega_k^2}-\frac{2\boldsymbol{\varphi}_k\zeta_k\dot{y}_k(t)}{\omega_k}-\frac{\boldsymbol{\varphi}_k\ddot{y}_k(t)}{\omega_k^2}\right) \qquad [3.10]$$

According to Besselink *et al.* [BES 13], the inverse of the stiffness matrix can be represented as

$$\mathbf{K}^{-1}=\sum_{k=1}^{n}\frac{\boldsymbol{\varphi}_k\boldsymbol{\varphi}_k^{\mathrm{T}}}{\omega_k^2} \qquad [3.11]$$

Notice that in MDM, $n$ is reduced to $l$ for all three terms in equation [3.10]. However in MAM, it is not necessary to approximate the first term in equation [3.10] using only $l$ modes, because this term can be exactly solved as a static problem by means of equation [3.11] to include all $n$ modes. According to equation [3.9], the second and third parts of equation [3.10] can be written as:

$$\sum_{k=1}^{l}\left(-\frac{2\boldsymbol{\varphi}_k\zeta_k\dot{y}_k(t)}{\omega_k}-\frac{\boldsymbol{\varphi}_k\ddot{y}_k(t)}{\omega_k^2}\right)=\sum_{k=1}^{l}\left(\boldsymbol{\varphi}_k y_k(t)-\frac{\boldsymbol{\varphi}_k\boldsymbol{\varphi}_k^{\mathrm{T}}\mathbf{f}(t)}{\omega_k^2}\right) \qquad [3.12]$$

Hence, based on the substitution of equation [3.11] into equation [3.10] and the combination with equation [3.12], the MAM results in the approximation of the displacement response:

$$\mathbf{u}(t)=\mathbf{K}^{-1}\mathbf{f}(t)+\sum_{k=1}^{l}\left(\boldsymbol{\varphi}_k y_k(t)-\frac{\boldsymbol{\varphi}_k\boldsymbol{\varphi}_k^{\mathrm{T}}\mathbf{f}(t)}{\omega_k^2}\right) \qquad [3.13]$$

The further introduction of equation [3.11] simplifies equation [3.13] into:

$$\mathbf{u}(t)=\sum_{k=1}^{l}\boldsymbol{\varphi}_k y_k(t)+\sum_{k=l+1}^{n}\frac{\boldsymbol{\varphi}_k\boldsymbol{\varphi}_k^{\mathrm{T}}}{\omega_k^2}\mathbf{f}(t)=\sum_{k=1}^{l}\boldsymbol{\varphi}_k H_k\boldsymbol{\varphi}_k^{\mathrm{T}}\mathbf{f}(t)+\sum_{k=l+1}^{n}\frac{\boldsymbol{\varphi}_k\boldsymbol{\varphi}_k^{\mathrm{T}}}{\omega_k^2}\mathbf{f}(t) \qquad [3.14]$$

The comparison between formulations of MDM and MAM indicates that the second term of equation [3.14] related to MAM could be treated as the correction term of MDM related to equation [3.8]. The computing error related to MDM increases, along with the increase in the number of DOFs. Obviously, more modes should be employed if MDM is used for large-scale problems. However, it is hard to decide how many modes should be employed in advance to obtain the prescribed accuracy. In the works of [BES 13] and [COR 83], detailed comparisons between the computing accuracies of MDM and MAM were made. It was concluded that the MAM outperformed the MDM in all cases. Here, their effects will be focused on topology optimization. Besides, other approximation methods such as the modal truncation augmentation method [DIC 97], Ritz vector method [WIL 82], quasi-static Ritz vector method [GU 00] and Pade approximation method [JEN 07] could also be employed to improve the accuracy of dynamic responses without increasing the computing cost prohibitively.

### 3.2.3. *Full method*

As an exact method of harmonic analysis, the FM is formulated by introducing the following solution form:

$$\mathbf{u}(t) = \mathbf{U}e^{\iota\omega t} \qquad [3.15]$$

where **U** is the complex amplitude vector of displacement response. Equation [3.1] can then be rewritten as:

$$(\mathbf{K} - \mathbf{M}\omega^2 + \iota\omega\mathbf{C})\mathbf{U} = \mathbf{F} \qquad [3.16]$$

**U** can directly be solved as

$$\mathbf{U} = (\mathbf{K} - \mathbf{M}\omega^2 + \iota\omega\mathbf{C})^{-1}\mathbf{F} \qquad [3.17]$$

in which Rayleigh damping corresponds to

$$\mathbf{C} = \mathrm{a}\mathbf{M} + \mathrm{b}\mathbf{K} \qquad [3.18]$$

Notice that constants $a$ and $b$ are Rayleigh damping coefficients of the structure. From the third relation of equation [3.2], the damping ratio $\zeta_k$ can be expressed as

$$\zeta_k = \frac{a + b\omega_k^2}{2\omega_k} \qquad [3.19]$$

Equation [3.16] can be solved using the same sparse direct solver as in a static analysis but with complex arithmetic by default. So the displacement response related to FM can be written as:

$$\mathbf{u}(t) = (\mathbf{K} - \mathbf{M}\omega^2 + \iota\omega\mathbf{C})^{-1}\mathbf{F}e^{\iota\omega t} \qquad [3.20]$$

### 3.2.4. Comparative tests of harmonic analysis methods

In this section, three numerical tests are studied to compare the above three analysis methods. The computing results obtained by FM are regarded as the exact solution in the comparison. In all numerical tests, the Young's modulus, Poisson ratio and density of the solid material are 200 GPa, 0.3 and 7,800 kg/m$^3$ respectively.

*Structure 1: 2D cantilever beam*

The structure is a rectangular domain of size 80 mm × 50 mm × 1 mm. It is clamped at the left side, as shown in Figure 3.1. Here, the domain is meshed into 48 × 30 plane stress elements with 3,038 DOFs in all. A harmonic force with the amplitude of 9 kN is applied at the middle node of the right edge.

**Figure 3.1.** *Structure 1: 2D cantilever beam (3,038 DOFs)*

*Structure 2: short 3D beam*

The beam structure has a size of 2 m × 1 m × 1 m. It is simply supported at the four corners of the bottom as shown in Figure 3.2. The design domain is meshed into 40 × 20 × 20 solid elements with 54,243 DOFs in all. A harmonic force with the amplitude of 1,000 kN is applied at the middle point of the tip face.

The same Rayleigh damping is adopted in all cases with $a = 10^{-3}$ and $b = 10^{-6}$. Here, MDM and MAM are all implemented with the first $l = 30$ modes for structures 1 and 2. Notice that structure 1 represents the common problem studied in the previous research with thousands of DOFs. Structure 2 (Figure 3.2) is used to illustrate large-scale problems of huge numbers of DOFs that really represent dynamic topology optimization from the engineering viewpoint.

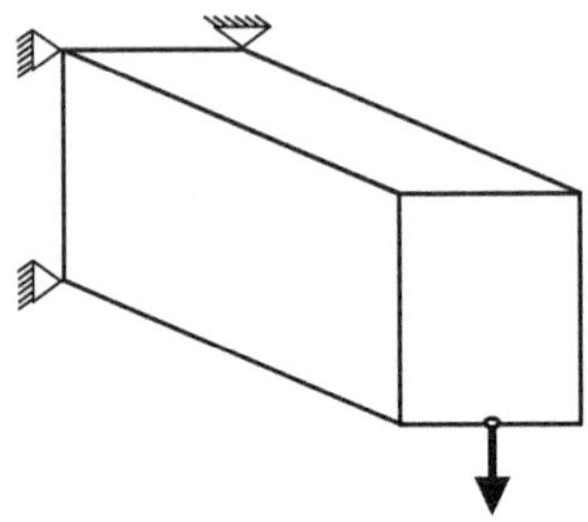

**Figure 3.2.** *Structure 2: 3D beam (99,603 DOFs)*

### 3.2.4.1. *Computing accuracy*

The displacement amplitude at the loaded point along the force direction is obtained by three different methods. They are compared in form of logarithm in Figures 3.3 and 3.4. Notice that the excitation frequency in each problem is always covered by the first $l$ eigenfrequencies selected in MDM and MAM by default.

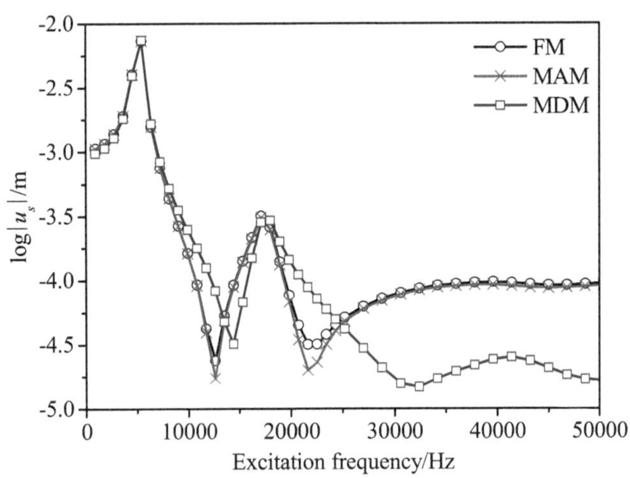


**Figure 3.3.** *Displacement amplitudes obtained by three methods for structure 1*

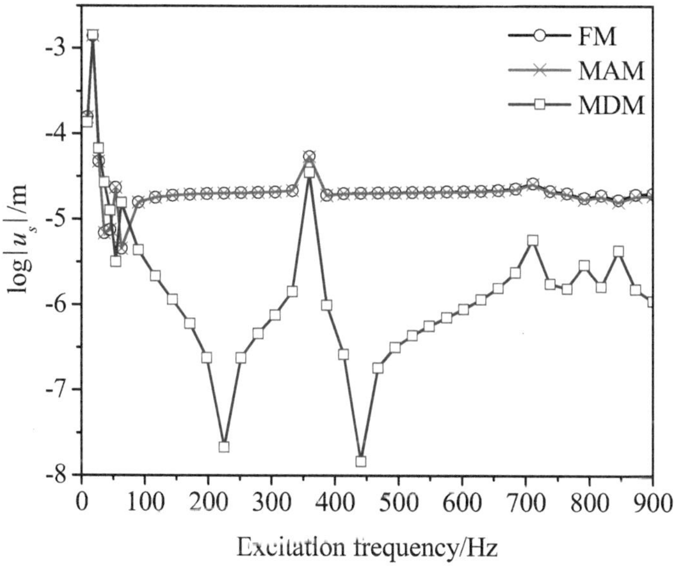


**Figure 3.4.** *Displacement amplitudes obtained by three methods for structure 2*

In both examples, we see that the displacement amplitude using the MAM is close to the exact solution obtained by the FM. Instead, the accuracy of the MDM is acceptable only when the excitation frequency is low. We also note that the error of MDM is relatively small near the peaks of response curves where the excitation frequency attains the resonant eigenfrequency.

Moreover, with the dramatic increase in the DOF number, the ratio of the employed mode numbers ($l = 30$) to $n$ greatly decreases in the MDM and MAM. Naturally, the accuracy of MDM and MAM would also decrease. To further confirm the above statement, structure 1 is analyzed using different meshes, while $l = 30$ is unchanged in the calculation of the harmonic response at the loading position. Here the excitation frequency is $f = 1{,}500$ Hz. The displacement amplitudes of structure 1 along with the mesh refinement are plotted in Figure 3.5.

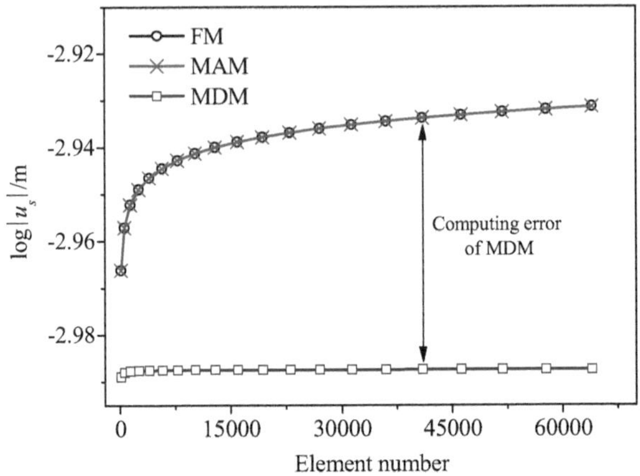


**Figure 3.5.** *Displacement response amplitudes versus mesh refinements for structure 1*

As is seen, both MAM and FM converge asymptotically along with the increase in element number or equivalently the number of DOFs. Comparatively, the computing error related to MDM increases along with the increase in the number of DOFs. Obviously, more modes should be employed if MDM is used for large-scale problems. However, it is hard to decide how many modes should be employed in advance to obtain the prescribed accuracy. Hence, FM and MAM outperform the MDM in computing accuracy, especially for large-scale problems.

### 3.2.4.2. *Computing efficiency*

Regarding the computing efficiency, the following two situations are considered.

#### *Case 1: Harmonic analysis at one specific excitation frequency*

Noticeably, the computing cost of FM is a little more than the static analysis due to the complex arithmetic. For MDM and MAM, most computing time is sacrificed to calculate modes while the mode superposition process is trivial. In fact, various methods exist to obtain the modes and detailed information about numerical algorithms can be found in [GRI 94, WIL 82]. Suppose that the computing time is $t_{FM} = \tau$

for FM under the harmonic force excitation with one specific frequency, and the computing time for MDM is $t_{\mathrm{MDM}}$ consisting of mode computing time $\delta$ and computing time $\lambda$ for the mode superposition process. Then,

$$\begin{aligned} t_{\mathrm{MDM}} &= \delta + \lambda \\ t_{\mathrm{FM}} &= \tau \end{aligned} \qquad [3.21]$$

As is well known, the mode computing time $\delta$ is usually greater than $\tau$, especially when the number of selected modes is large with $t_{\mathrm{MDM}} >> t_{\mathrm{FM}}$. In order to ascertain the above conclusion, structure 1 is used to test the efficiency of MDM and FM. All the calculations are implemented in ANSYS software with a mesh of 320 × 200 plane stress elements, and the Block Lanczos method is adopted to proceed the modal analysis. Here, the specific excitation frequency is 1,500 Hz. Finally, we have $t_{\mathrm{FM}}$ = 8.53s, while $t_{\mathrm{MDM}}$ depends upon the number of included modes $l$, as shown in Figure 3.6. Meanwhile, the computing time $\delta$ for the modal analysis extracting $l$ modes is also plotted.

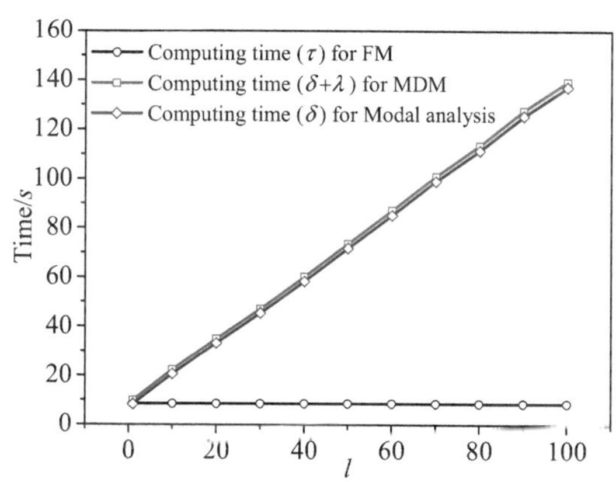


**Figure 3.6.** *Computing time comparisons of FM, MDM and modal analysis*

In this case, it can be seen that modal analysis takes about the whole time in the MDM with $t_{\mathrm{MDM}} \approx \delta$ and that the latter is always larger than $t_{\mathrm{FM}}$ even when only one mode is used. $t_{\mathrm{MDM}}$ approximately increases in a linear way with respect to $l$. Oppositely, little time ($\lambda = 2.1$ s) is consumed for the mode superposition process in the case of $l$ = 100. FM demonstrates its great advantage in both computing efficiency and accuracy for the harmonic analysis with one specific frequency.

*Case 2: Harmonic analysis at multiple excitation frequencies*

Suppose $g$ is the number of sampling excitation frequencies. The computing time can then be estimated as:

$$\begin{aligned} t_{\mathrm{MDM}} &= \delta + g\lambda \\ t_{\mathrm{FM}} &= g\tau \end{aligned} \qquad [3.22]$$

Each excitation frequency implies that one independent complex analysis is required if FM is used. Comparatively, the computing cost related to MDM and MAM would not increase greatly because only one modal analysis is needed for a given structure and the cost of superposition process $\lambda$ is trivial. Therefore, $t_{\mathrm{FM}} >> t_{\mathrm{MDM}}$ and $t_{\mathrm{FM}} >> t_{\mathrm{MAM}}$ hold if $g$ is large enough. This is why MDM is widely used in practice because hundreds even thousands of excitation frequencies may be sampled to capture the corresponding displacement amplitudes. But considering both the efficiency and the accuracy, the MAM would be the best choice for the harmonic problems in this situation.

## 3.3. Topology optimization under harmonic force excitations

### 3.3.1. *Topology optimization formulation*

For a dynamic problem under harmonic force excitations, topology optimization formulation is often stated as:

$$\begin{array}{ll} \text{find} & \boldsymbol{\eta} = \{\eta_i\} \quad i = 1,\ldots,n_{\mathrm{e}} \\ \min & |u_s(t)| \\ \text{s.t} & V \le V_{\mathrm{U}} \\ & 0 < \eta_{\mathrm{L}} \le \eta_i \le 1 \end{array} \qquad [3.23]$$

where $\eta_L$ is the lower bound of the set of design variables defined by element pseudo-densities. Here, $\eta_L = 0.001$ is used to prevent the mass, stiffness and damping matrices from becoming singular. $i$ and $V$ denote element number and the solid volume fraction, respectively. $V_U$ is the upper bound of the latter.

In topology optimization of dynamic problems, it is recognized that the SIMP interpolation scheme would cause localized modes phenomena for the stiffness and mass matrices, because of the mismatch between element stiffness and mass. This has been discussed in detail in Chapter 2 and the polynomial interpolation model is adopted here.

### 3.3.2. *Sensitivity analysis*

For a harmonic problem, the sensitivity of displacement with respect to pseudo-density variables $\eta_i$ is introduced in this section. Using MDM or MAM, the sensitivity of displacement can be obtained by directly differentiating equation [3.8] or equation [3.13]. Thus, the MDM corresponds to:

$$\frac{\partial \mathbf{u}(t)}{\partial \eta_i} = \sum_{k=1}^{l} \left( \frac{\partial \boldsymbol{\varphi}_k}{\partial \eta_i} H_k \boldsymbol{\varphi}_k^{\mathrm{T}} + \boldsymbol{\varphi}_k \frac{\partial H_k}{\partial \eta_i} \boldsymbol{\varphi}_k^{\mathrm{T}} + \boldsymbol{\varphi}_k H_k \frac{\partial \boldsymbol{\varphi}_k^{\mathrm{T}}}{\partial \eta_i} \right) \mathbf{f}(t) \qquad [3.24]$$

Suppose:

$$Z_k = (\omega_k^2 - \omega^2 + 2\iota\zeta_k \omega_k \omega)^{-1} - \omega_k^{-2} \qquad [3.25]$$

The sensitivity of displacement using MAM can be written as:

$$\frac{\partial \mathbf{u}(t)}{\partial \eta_i} = \frac{\partial \left( \mathbf{K}^{-1} \mathbf{F} \right)}{\partial \eta_i} e^{\iota \omega t} + \sum_{k=1}^{l} \left( \frac{\partial \boldsymbol{\varphi}_k}{\partial \eta_i} Z_k \boldsymbol{\varphi}_k^{\mathrm{T}} + \boldsymbol{\varphi}_k \frac{\partial Z_k}{\partial \eta_i} \boldsymbol{\varphi}_k^{\mathrm{T}} + \boldsymbol{\varphi}_k Z_k \frac{\partial \boldsymbol{\varphi}_k^{\mathrm{T}}}{\partial \eta_i} \right) \mathbf{f}(t) \qquad [3.26]$$

Obviously, sensitivities of eigenfrequencies and eigenvectors are basic calculations for solutions of equation [3.24] and equation [3.26].

The sensitivity of eigenfrequency can be obtained by the following equation:

$$\frac{\partial \omega_k}{\partial \eta_i} = \frac{\boldsymbol{\varphi}_k^{\mathrm{T}} \left( \frac{\partial \mathbf{K}}{\partial \eta_i} - \omega_k^2 \frac{\partial \mathbf{M}}{\partial \eta_i} \right) \boldsymbol{\varphi}_k}{2\omega_k} \quad [3.27]$$

The derivatives of the eigenvectors hold the following form [ALV 97]:

$$\frac{\partial \boldsymbol{\varphi}_k}{\partial \eta_i} = \sum_{r=1}^{l} B_{kr} \boldsymbol{\varphi}_r \quad [3.28]$$

where $B_{kr}$ is calculated as:

$$B_{kr} = \begin{cases} \dfrac{\boldsymbol{\varphi}_r^{\mathrm{T}} \left( \dfrac{\partial \mathbf{K}}{\partial \eta_i} - \dfrac{\partial \omega_k^2}{\partial \eta_i} \mathbf{M} - \omega_k^2 \dfrac{\partial \mathbf{M}}{\partial \eta_i} \right) \varphi_k}{\omega_r^2 - \omega_k^2} & r \neq k \\ -\dfrac{1}{2} \boldsymbol{\varphi}_k^{\mathrm{T}} \dfrac{\partial \mathbf{M}}{\partial \eta_i} \boldsymbol{\varphi}_k & r = k \end{cases} \quad [3.29]$$

Suppose $\boldsymbol{\Lambda}$ is the column vector with all terms being zero except term $s$ being 1. So the sensitivity of displacement of concerned DOF $s$ can be calculated as:

$$\frac{\partial u_s(t)}{\partial \eta_i} = \boldsymbol{\Lambda}^{\mathrm{T}} \frac{\partial \mathbf{u}(t)}{\partial \eta_i} \quad [3.30]$$

As for the displacement response using FM, the differentiation of equation [3.16] gives rise to

$$\frac{\partial \mathbf{K}_d}{\partial \eta_i} \mathbf{U} + \mathbf{K}_d \frac{\partial \mathbf{U}}{\partial \eta_i} = \mathbf{0} \quad [3.31]$$

where the following notation is used:

$$\mathbf{K}_d = (\mathbf{K} - \mathbf{M}\omega^2 + \iota\omega\mathbf{C}) \qquad [3.32]$$

Suppose

$$u_s = \mathbf{\Lambda}^{\mathrm{T}}\mathbf{U} \qquad [3.33]$$

By virtue of the adjoint method, the following equation is then established:

$$\begin{aligned}\frac{\partial u_s}{\partial \eta_i} &= \mathbf{\Lambda}^{\mathrm{T}}\frac{\partial \mathbf{U}}{\partial \eta_i} - \boldsymbol{\lambda}^{\mathrm{T}}\left(\frac{\partial \mathbf{K}_d}{\partial \eta_i}\mathbf{U} + \mathbf{K}_d\frac{\partial \mathbf{U}}{\partial \eta_i}\right) \\ &= (\mathbf{\Lambda}^{\mathrm{T}} - \boldsymbol{\lambda}^{\mathrm{T}}\mathbf{K}_d)\frac{\partial \mathbf{U}}{\partial \eta_i} - \boldsymbol{\lambda}^{\mathrm{T}}\frac{\partial \mathbf{K}_d}{\partial \eta_i}\mathbf{U}\end{aligned} \qquad [3.34]$$

where $\boldsymbol{\lambda}$ is the adjoint vector obtained by

$$\mathbf{K}_d\boldsymbol{\lambda} = \mathbf{\Lambda} \qquad [3.35]$$

As a result, equation [3.34] can be written as:

$$\frac{\partial u_s}{\partial \eta_i} = -\boldsymbol{\lambda}^{\mathrm{T}}\frac{\partial \mathbf{K}_d}{\partial \eta_i}\mathbf{U} \qquad [3.36]$$

After the sensitivity of displacement under harmonic force excitation is obtained by means of MDM, MAM and FM, the sensitivity of displacement amplitude can be derived through the chain rule as:

$$\frac{\partial |u_s(t)|}{\partial \eta_i} = \left( real(u_s)\cdot real\left(\frac{\partial u_s}{\partial \eta_i}\right) + imag(u_s)\cdot imag\left(\frac{\partial u_s}{\partial \eta_i}\right) \right)|u_s(t)|^{-1} \qquad [3.37]$$

### 3.3.3. *Numerical examples*

In this section, topology optimization problems under harmonic force excitation at one specific frequency and multiple frequencies are solved. The GCMMA algorithm is used as the optimizer.

#### 3.3.3.1. *Harmonic excitation at one specific frequency*

*– 2D examples*

The structure is illustrated in Figure 3.1. Suppose the vertical displacement amplitude at the loading position is minimized. The volume fraction of solid material is constrained to be less than 50%. Initial values of all pseudo-densities are set to be 0.5 with the first eigenfrequency of the initial structure being 1,776 Hz, correspondingly. At the specific excitation frequency of $f = 1{,}500$ Hz, $l = 30$ modes are employed by MDM and MAM. The optimized configurations are shown in Figure 3.7(a) and the iteration curves of the objective function are plotted in Figure 3.7(b).

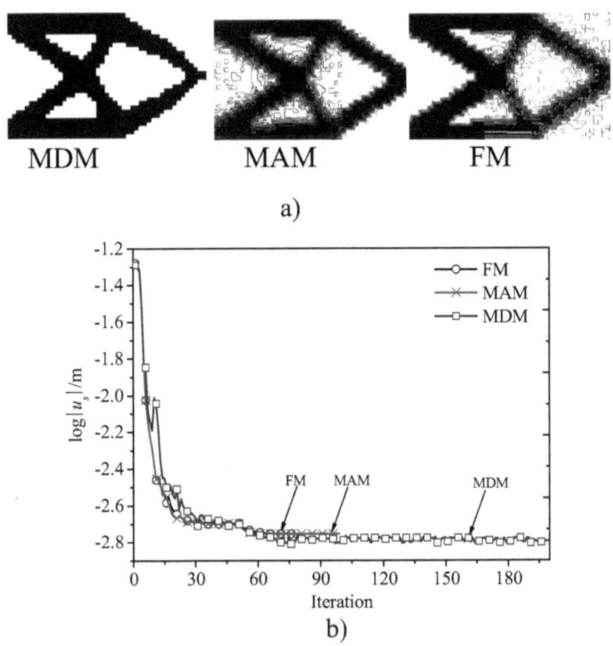


**Figure 3.7.** *Optimization results of structure 1 with a mesh of 48 × 30 (3,038 DOFs): a) optimized configurations; b) iteration curves of the objective function*

It can be seen that nearly the same configurations are obtained by three methods. As the excitation frequency $f = 1{,}500$ Hz is lower than the initial first eigenfrequency (1,776 Hz), solutions are comparable with that minimizing the static compliance when a static load is applied. Different problems with the similar order of DOFs can also be found in references.

Therefore, all three methods are effective in this case. Now, the structure in Figure 3.1 is further optimized with refined meshes of large numbers of DOFs. Figures 3.8 and 3.9 show the topology optimization results for a mesh of 160 × 100 and 240 × 150 elements, respectively. While $l$ = 30, modes are still employed in MDM and MAM. In conclusion, a poor convergence occurs for MDM. The optimized configuration is not clear, especially near the loading point.

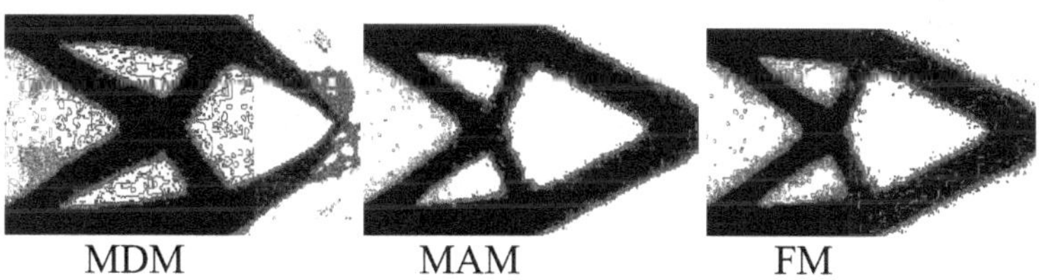


**Figure 3.8.** *Optimized configurations of structure 1 with a mesh of 160 × 100 (32,522 DOFs)*

**Figure 3.9.** *Optimized configurations of structure 1 with a mesh of 240 × 150 (72,782 DOFs)*

Examples show that MAM and FM work well in all the above cases, while MDM converges poorly due to the low accuracy of harmonic responses when large-scale problems are considered. The reason can be explained in the following two aspects.

First, as a kind of approximate method, MDM would introduce errors because of the truncation modes. Although more modes may reduce the error, it is difficult to decide how many modes should be employed to ensure the accuracy of dynamic solutions in advance. Meanwhile, as the structure is modified in each iteration of optimization, it is difficult to choose a proper $l$ instantaneously. The best way is to deal with the problem by selecting the value of $l$ large enough for all iterations, which means a huge computing cost.

Second, as mentioned in section 3.2.4.1, the large error of MDM occurs when the excitation frequency is far from the resonant eigenfrequencies. The dynamic response minimization aims at pushing the eigenfrequencies far away from the excitation frequency , which will further deteriorate the accuracy of harmonic response for the prescribed excitation frequency. This might explain why the MDM leads to the incorrect solution for large-scale problems. It should be considered that this problem could be avoided at the cost of huge amounts of computing by employing enough numbers of modes in MDM. Therefore, MDM is preferable to deal with small-scale problems.

– *3D examples*

A literature review indicates that topology optimizations of large-scale dynamic problems are rarely reported due to the prohibitive computing time and convergence difficulty related to the MDM. To further verify the effects of three analysis methods within this context, consider now 3D structures of huge numbers of DOFs. Suppose the displacement amplitude at the loading position along the force direction is minimized. For all tests below, $l = 30$ modes are employed in MDM and MAM. The volume fraction is constrained to be less than 10% and the initial values of all pseudo-densities are uniformly set to be 0.1 correspondingly.

The 3D beam structure illustrated in Figure 3.2 is also studied. The first eigenfrequency of the initial structure is 2.95 Hz. Optimization results with excitation frequency $f = 2.5$ Hz are shown in Figure 3.10(a) and the iteration histories of the objective function are also presented in Figure 3.10(b).

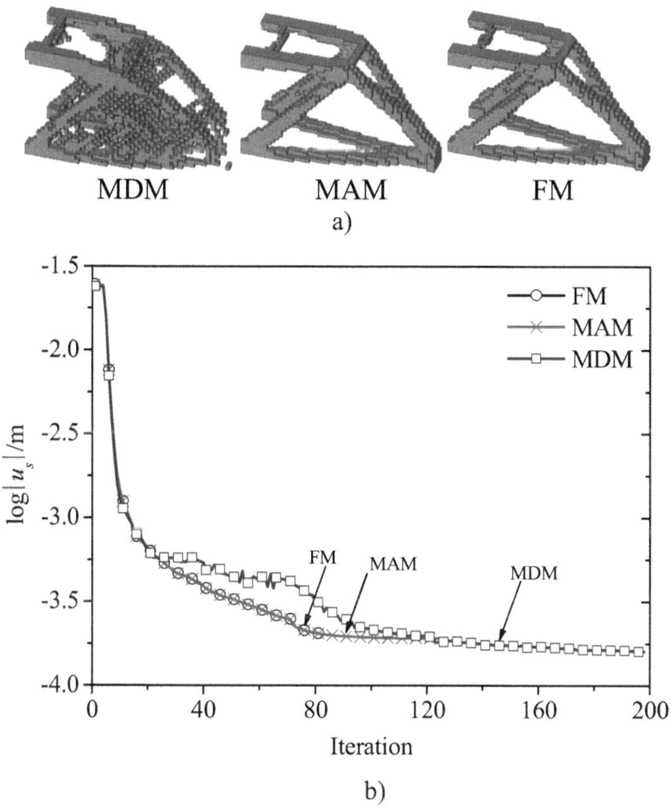


**Figure 3.10.** *Optimization results of structure 2 (72,782 DOFs): a) optimized configurations; b) iteration curves of objective function*

It can be seen from all the above numerical tests that the final design configuration with MDM becomes unreasonable with the increase in DOFs, while MAM or FM can ensure the optimized configurations of the structure due to their high accuracy. The iteration curves of objective function also illustrate that FM is the most powerful in convergence.

### 3.3.3.2. *Harmonic excitations with multiple frequencies*

In topology optimization related to harmonic responses, a converged solution could be easily obtained if the excitation frequency is lower than the resonant eigenfrequency, as shown in section 3.3.3.1. However, in the case of single excitation frequency with a higher value than the resonant eigenfrequency, the optimization iteration would be difficult to converge. Meanwhile, the parasitic effect of the material layout is mostly evident in the obtained configurations and the static stiffness is usually quite weak. This phenomenon was also mentioned by Olhoff and Du [OLH 05]. Actually, practical structures are usually excited by the harmonic load in a frequency interval, not just at a prescribed frequency value. Meanwhile, the static stiffness is also an essential requirement. As is well known, the minimization of dynamic response under one low frequency excitation is consistent with the maximization of static stiffness, to some extent. Therefore, in order to achieve practical optimized configurations, two schemes are adopted in structural design. One is to introduce the static compliance into the optimization formulation as an additional constraint, as detailed by Olhoff and Du [OLH 05]. The other scheme is to handle the harmonic excitations with multiple frequencies or in frequency intervals, which will be presented below.

For the optimization problem under harmonic excitations with multiple frequencies, the integral of displacement amplitude in a frequency interval $[\omega_A, \omega_B]$ is usually considered as the objective function. The optimization formulation can be stated as

$$\begin{aligned} &\text{find} && \boldsymbol{\eta} = \{\eta_i\} \quad i = 1,2,\ldots,n_{\mathrm{e}} \\ &\min && \int_{\omega_A}^{\omega_B} \left| u_s(\omega) \right| d\omega \\ &\text{s.t} && V \le V_{\mathrm{U}} \\ &&& 0 < \eta_{\mathrm{L}} \le \eta_i \le 1 \end{aligned} \qquad [3.38]$$

The calculation of the integral in a frequency interval is crucial to the optimization. As is well known, the curve of harmonic response demonstrates a very sharp jump around the resonant eigenfrequency. It therefore takes a significant effort to yield converged solutions

[CHE 98]. In this section, the numerical integration method, i.e. Gauss–Legendre integration, is used to calculate the integral:

$$\int_{\omega_A}^{\omega_B} \left|u_s(\omega)\right| d\omega \approx \frac{\omega_B - \omega_A}{2} \sum_{\varsigma=1}^{N} \gamma_\varsigma \left| x_s\left(\frac{\omega_B - \omega_A}{2} v_\varsigma + \frac{\omega_B + \omega_A}{2}\right) \right| \qquad [3.39]$$

where $\gamma_\varsigma$ is the weight factor for the ςth Gaussian-point, $v_\varsigma$ is the Gaussian-point within [–1,1] and $N$ is the number of Gaussian-points.

When the integrand is very complicated or the integration interval is very large, the subdivision of the integration interval is necessary to ensure the computing accuracy. Considering the sharp jumps of harmonic response curve and the large frequency interval, the latter is subdivided by the eigenfrequencies first. Each subinterval between adjacent eigenfrequencies will be further subdivided by $m$ additional points. The distribution of the $j$th point within the subinterval [$\omega_k$, $\omega_{k+1}$] is then defined by

$$\omega_{k,j} = \omega_k + w_j\left(\omega_{k+1} - \omega_k\right) \quad 1 \le j \le m, \quad 0 < w_j < 1 \qquad [3.40]$$

where $w_j$ is the proportional factor of the $j$th point in the subinterval. In order to represent the sharp jumps, $w_j$ should be selected properly to form small subintervals near the eigenfrequency and large subintervals far from the eigenfrequency. In this chapter, $m = 5$ additional points are chosen in the following way between adjacent eigenfrequencies:

$$\begin{aligned}
\omega_{k,1} &= \omega_k + 0.01\left(\omega_{k+1} - \omega_k\right) \\
\omega_{k,2} &= \omega_k + 0.11\left(\omega_{k+1} - \omega_k\right) \\
\omega_{k,3} &= \omega_k + 0.5\left(\omega_{k+1} - \omega_k\right) \\
\omega_{k,4} &= \omega_k + 0.89\left(\omega_{k+1} - \omega_k\right) \\
\omega_{k,5} &= \omega_k + 0.99\left(\omega_{k+1} - \omega_k\right)
\end{aligned} \qquad [3.41]$$

Figure 3.11 illustrates the subdivision of the frequency interval.

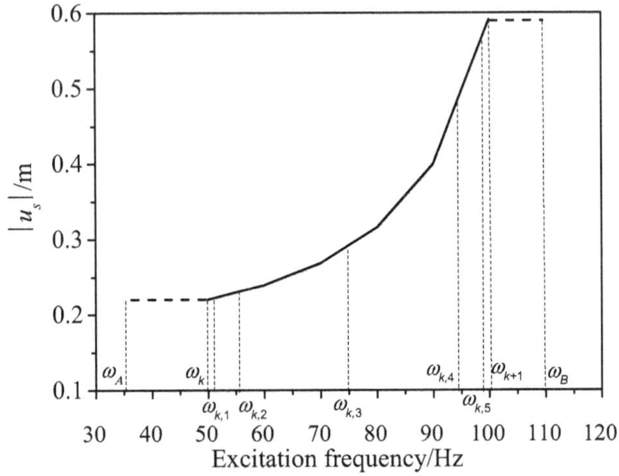


**Figure 3.11.** *Schematic of the frequency interval subdivision*

After the subdivision, each subinterval would be calculated by the Gauss–Legendre integration method with 15 Gaussian-points. Examples show that this method works well in the optimization to deal with harmonic responses with hundreds of excitation frequencies. Optimization with FM is too time-consuming to implement and hence only MDM and MAM are adopted in this section.

The 3D cantilever beam has a size of 0.8 m × 0.4 m × 0.06 m and is clamped at the left side, as shown in Figure 3.12. The design domain is meshed into 80 × 40 × 6 solid elements. A harmonic force with the amplitude of 1,000 kN is applied at the center of the right side. The integral of vertical displacement amplitude at the loading position is minimized and the volume fraction is constrained to be less than 30%. Initial values of all pseudo-densities are set to be 0.3. Correspondingly, the first eigenfrequency of the initial structure is 21 Hz.

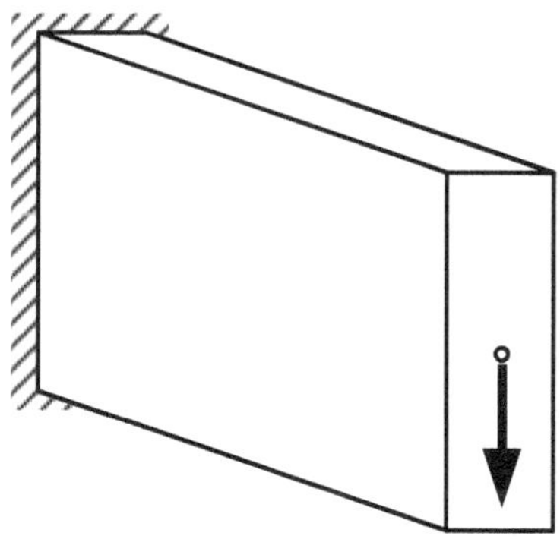

**Figure 3.12.** *Structure 3: 3D cantilever beam (69,741 DOFs)*

Three frequency intervals are considered here, namely, [0–100] Hz, [0–300] Hz and [0–900] Hz. Notice that only MDM and MAM are adopted due to the prohibitive computing time of the FM. Here, $l = 15$ modes are employed in MDM and MAM for the first two frequency intervals, while $l = 20$ modes are employed for the last interval. Optimized configurations with MDM and MAM are shown in Figures 3.13 and 3.14, respectively. Obviously, the convergence is very poor with MDM, while the configurations by MAM are quite clear. For the optimized configurations in Figure 3.14, the response curves calculated using FM are shown in Figure 3.15. It can be clearly seen that the responses in the frequency intervals decrease.

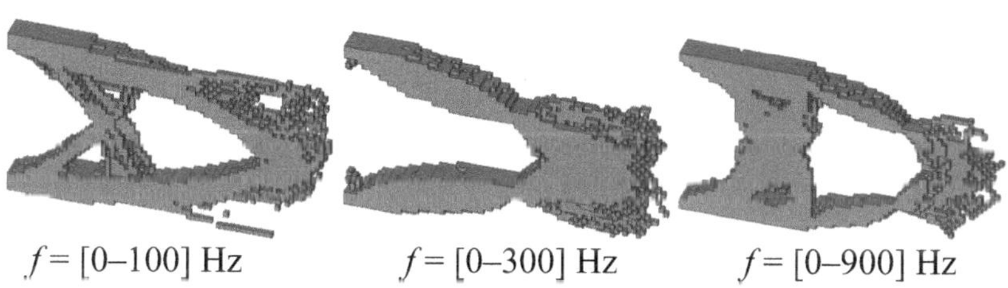


**Figure 3.13.** *Optimized configurations of structure 3 using MDM*

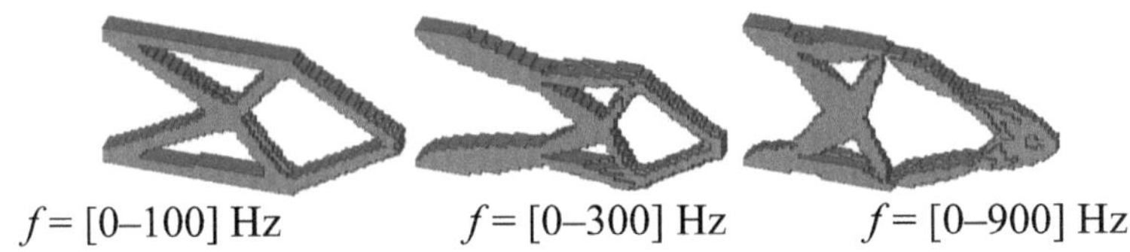


**Figure 3.14.** *Optimized configurations of structure 3 using MAM*

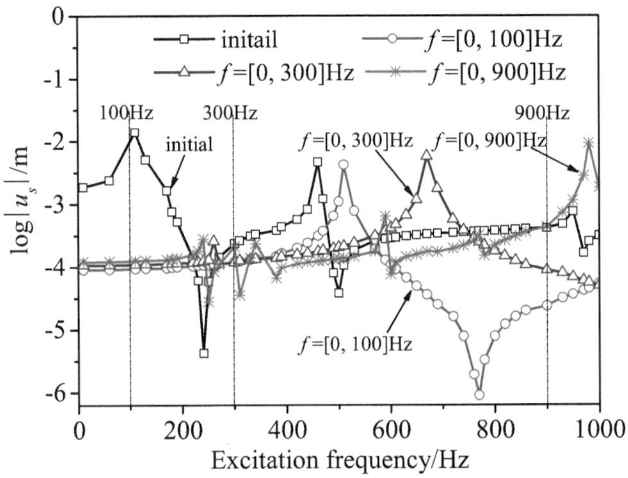


**Figure 3.15.** *Displacement amplitudes of initial and three optimized structures using FM. For a color version of this figure, see www.iste.co.uk/zhang/topology.zip*

Numerical tests find that the MAM is the most favorable method, considering the balance between computing accuracy and efficiency, for the optimization problem under harmonic force excitations with multiple excitation frequencies. Comparatively, the computing accuracy of the MDM makes the optimization process difficult to converge. The computing cost of the FM is unacceptable in the case of harmonic excitations with multiple frequencies.

## 3.4. Analysis methods for stationary random force excitations

### 3.4.1. *Complete quadratic combination method*

Now, consider a discretized $n$-DOF structure subjected to stationary random force excitation. The motion equation can be written as:

$$\mathbf{M}\ddot{\mathbf{u}}(t)+\mathbf{C}\dot{\mathbf{u}}(t)+\mathbf{K}\mathbf{u}(t)=\mathbf{b}\mathbf{p}(t) \quad [3.42]$$

$\mathbf{p}(t)$ is a $d$-dimension stationary random force vector of non-zero values, whose power spectral density (PSD) matrix is of $d$-dimension and denoted by $\mathbf{S}_p(\omega)$. Notice that $\mathbf{b}$ is a $n \times d$ transformation matrix representing the force distribution. Here, $\mathbf{bp}(t)$ is assumed to be white-noise excitation with zero mean value, i.e. a uniform power spectral density over the frequency interval.

As discussed in section 3.2.1, a number of $n$ uncoupled equations of motion can be obtained.

$$\ddot{y}_k(t)+2\zeta_k\omega_k\dot{y}_k(t)+\omega_k^2 y_k(t)=\boldsymbol{\varphi}_k^{\mathrm{T}}\mathbf{b}\mathbf{p}(t) \quad [3.43]$$

By means of the Duhamel integral, the time-domain solution of this equation is [CLO 75]:

$$y_k(t)=\int_{-\infty}^{\infty}\boldsymbol{\varphi}_k^{\mathrm{T}}\mathbf{b}\mathbf{p}(t-\tau)h_k(\tau)d\tau \quad [3.44]$$

where $h_k(\tau)$ is the unit impulse response function related to the single DOF system of equation [3.43]:

$$h_k(\tau)=\begin{cases}\dfrac{e^{\xi_k\omega_k\tau}}{m_k\omega_k}\sin\omega_k\tau & \tau\geq 0\\ 0 & \tau<0\end{cases} \quad [3.45]$$

with

$$m_k = \boldsymbol{\varphi}_k^{\mathrm{T}} \mathbf{M} \boldsymbol{\varphi}_k \quad [3.46]$$

It follows that:

$$\mathbf{u}(t) = \sum_{k=1}^{n} \boldsymbol{\varphi}_k y_k = \sum_{k=1}^{n} \boldsymbol{\varphi}_k \boldsymbol{\varphi}_k^{\mathrm{T}} \mathbf{b} \int_{-\infty}^{\infty} \mathbf{p}(t-\tau) h_k(\tau) d\tau \quad [3.47]$$

The autocorrelation function of displacement response $\mathbf{u}(t)$ reads [CLO 75]:

$$\begin{aligned} \mathbf{R}_u(\Delta) &= \mathrm{E}[\mathbf{u}(t)\mathbf{u}(t+\Delta)^{\mathrm{T}}] \\ &= \sum_{k=1}^{n} \sum_{j=1}^{n} \boldsymbol{\varphi}_k \boldsymbol{\varphi}_k^{\mathrm{T}} \mathbf{b} \left( \int_{-\infty}^{\infty} \int_{-\infty}^{\infty} \mathbf{R}_p(\Delta + \tau_1 - \tau_2) h_k(\tau_1) h_j(\tau_2) d\tau_1 d\tau_2 \right) \mathbf{b}^{\mathrm{T}} \boldsymbol{\varphi}_j \boldsymbol{\varphi}_j^{\mathrm{T}} \end{aligned} \quad [3.48]$$

The PSD matrix of random displacement response $\mathbf{S}_u(\omega)$ can then be obtained by Fourier transformation of the above autocorrelation function:

$$\mathbf{S}_u(\omega) = \frac{1}{2\pi} \int_{-\infty}^{\infty} \mathbf{R}_u(\Delta) e^{-\iota\omega\Delta} \mathrm{d}\Delta = \sum_{k=1}^{n} \sum_{j=1}^{n} H_k^* H_j \boldsymbol{\varphi}_k \boldsymbol{\varphi}_k^{\mathrm{T}} \mathbf{b} \mathbf{S}_p(\omega) \mathbf{b}^{\mathrm{T}} \boldsymbol{\varphi}_j \boldsymbol{\varphi}_j^{\mathrm{T}} \quad [3.49]$$

Actually, it is almost impossible to use all the $n$ modes in the computing process of equation [3.49], especially for large-scale problems. Suppose $l$ is the number of modes employed in the computing with $l << n$, equation [3.49] is then approximated as:

$$\mathbf{S}_u(\omega) = \sum_{k=1}^{l} \sum_{j=1}^{l} H_k^* H_j \boldsymbol{\varphi}_k \boldsymbol{\varphi}_k^{\mathrm{T}} \mathbf{b} \mathbf{S}_p(\omega) \mathbf{b}^{\mathrm{T}} \boldsymbol{\varphi}_j \boldsymbol{\varphi}_j^{\mathrm{T}} \quad [3.50]$$

The CQC method [CLO 75, LIN 92] consists of obtaining the PSD matrix of random displacement response by computing equation [3.50] directly. Since the latter involves the cross-correlation terms between all $l$ participant modes, the computing would be very expensive for large values of $l$.

### 3.4.2. Conventional pseudo-excitation method

Since the PSD matrix $\mathbf{S}_p(\omega)$ is Hermitian, it can be decomposed into [LIN 01]:

$$\mathbf{S}_p(\omega)=\sum_{q=1}^{Q}(\boldsymbol{\gamma}_q)^*(\boldsymbol{\gamma}_q)^{\mathrm{T}} \qquad [3.51]$$

in which $Q$ is the rank of $\mathbf{S}_p(\omega)$. Therefore, equation [3.50] can be rewritten as:

$$\mathbf{S}_u(\omega)=\sum_{q=1}^{Q}\left(\sum_{j=1}^{l}\boldsymbol{\varphi}_j\boldsymbol{\varphi}_j^{\mathrm{T}}H_j\mathbf{b}\boldsymbol{\gamma}_q\right)^*\left(\sum_{k=1}^{l}\boldsymbol{\varphi}_k\boldsymbol{\varphi}_k^{\mathrm{T}}H_k\mathbf{b}\boldsymbol{\gamma}_q\right)^{\mathrm{T}} \qquad [3.52]$$

Suppose

$$\mathbf{g}_q(t)=\sum_{k=1}^{l}\boldsymbol{\varphi}_kH_k\boldsymbol{\varphi}_k^{\mathrm{T}}\mathbf{b}\boldsymbol{\gamma}_qe^{\imath\omega t} \qquad [3.53]$$

According to the MDM, $\mathbf{g}_q(t)$ is the displacement response vector of equation [3.42] under the $q$th pseudo harmonic force vector $\mathbf{b}\boldsymbol{\gamma}_qe^{\imath\omega t}$. Equation [3.52] can then be rewritten as:

$$\mathbf{S}_u(\omega)=\sum_{q=1}^{Q}\mathbf{g}_q^*\mathbf{g}_q^{\mathrm{T}} \qquad [3.54]$$

This is the conventional PEM [LIN 92, LIN 01]. It holds exactly the same accuracy as the CQC but is more efficient than the CQC for random vibration analysis. This method means that the PSD matrix of random displacement response can be solved through the harmonic responses under pseudo-harmonic excitations. To have a clear idea, suppose all the vectors and matrixes in equation [3.50], equation [3.53] and equation [3.54] are known in advance. The computing of equation [3.50] related to the CQC implies $n^2(3l^2 + 2dl^2) + nd^2l^2$ multiplication operations of real numbers, while equation [3.53] and equation [3.54] related to the PEM imply $n^2Q + n(2l + dl)Q$ multiplication operations. Generally, as $n$ is great in comparison with

$l$, $d$ and $Q$ ($Q \leqslant d$), the coefficient of $n^2$ for the PEM is therefore much smaller than that for the CQC.

### 3.4.3. *The combined method of PEM and MAM*

Other than the computing efficiency, the accuracy of random response is also very crucial to topology optimization. In fact, both the CQC and PEM belong to the same kind of mode superposition method and the truncation modes would undoubtedly introduce computing errors into the random responses. To remedy this, one possibility for the CQC so far is to increase the number of modes at the expense of huge computing cost especially for large-scale problems. Notably, this would make the dynamic topology optimization impractical. Meanwhile, it is difficult to decide how many modes should be employed in advance to guarantee the accuracy for a specified structure. Therefore, adopting a great number of modes is neither a reasonable nor practical strategy.

With the help of the conventional PEM, random responses are now solved through the pseudo-harmonic responses. This implies that the improvement of random response accuracy can be achieved by increasing the accuracy of pseudo-harmonic responses. Equation [3.53] is the so-called MDM commonly used in solving the harmonic responses due to its simplicity from the engineering viewpoint. However, the truncation of $n$-$l$ high-order modes would decrease the accuracy of response. Shi *et al.* [SHI 11] introduced MAM to replace MDM involved in the conventional PEM for structural analysis under multisupport excitations. As discussed below, the combined method of PEM and MAM makes it possible to improve the accuracy of random response with high efficiency. Generally, the MAM has a higher accuracy than the MDM in the case of using the same number of modes and can easily be implemented. Detailed comparisons between the computing accuracies of both methods have been made [COR 83, BES 13]. It was concluded that the MAM outperformed the MDM in all cases. The error of truncation modes in equation [3.53] related to the MDM can be remedied by the MAM using a pseudo-static solution. In this sense, equation [3.53] reads:

$$\mathbf{g}_q(t) = \left( \mathbf{K}^{-1}\left(\mathbf{b}\boldsymbol{\gamma}_q\right) + \sum_{k=1}^{l} (\boldsymbol{\varphi}_k H_k \boldsymbol{\varphi}_k^{\mathrm{T}} - \frac{\boldsymbol{\varphi}_k \boldsymbol{\varphi}_k^{\mathrm{T}}}{\omega_k^2}) \mathbf{b}\boldsymbol{\gamma}_q \right) e^{\iota\omega t} \qquad [3.55]$$

where $\mathbf{K}^{-1}(\mathbf{b}\boldsymbol{\gamma}_q)$ is the pseudo-static displacement under static force vector $(\mathbf{b}\boldsymbol{\gamma}_q)$. As equation [3.55] just needs one more additional static analysis than equation [3.53], efficiencies of MAM and MDM would be very close. Equation [3.55] can also be written as [BES 13]:

$$\mathbf{g}_q(t) = \sum_{k=1}^{l} \boldsymbol{\varphi}_k H_k \boldsymbol{\varphi}_k^{\mathrm{T}} \mathbf{b}\boldsymbol{\gamma}_q e^{\iota\omega t} + \sum_{k=l+1}^{n} \frac{\boldsymbol{\varphi}_k \boldsymbol{\varphi}_k^{\mathrm{T}}}{\omega_k^2} \mathbf{b}\boldsymbol{\gamma}_q e^{\iota\omega t} \qquad [3.56]$$

The second term of equation [3.56] related to MAM could be treated as the correction term to equation [3.53] of MDM. Thus, the accuracy of random response can be improved significantly by replacing equation [3.53] with equation [3.56].

In order to compare the accuracy between the conventional PEM involving the MDM and the combined method of PEM and MAM, the exact solution of PSD obtained by means of the full method (FM) is used as the benchmark for comparison. As an exact method of harmonic analysis, the FM directly solves the displacement response vector dominated by equation [3.42] under the $q$th pseudo-harmonic force vector $\mathbf{b}\boldsymbol{\gamma}_q e^{\iota\omega t}$.

$$\mathbf{g}_q(t) = (\mathbf{K} - \mathbf{M}\omega^2 + \iota\omega\mathbf{C})^{-1} \mathbf{b}\boldsymbol{\gamma}_q e^{\iota\omega t} \qquad [3.57]$$

The Rayleigh damping is also adopted as presented in section 3.2.3. Equation [3.57] can be handled by sparse direct solver as used in static analysis, but with complex arithmetic by default. By replacing equation [3.53] with equation [3.57], the PSD of random response would be the exact solution and will be used as the benchmark in the following comparisons. Notice that each excitation frequency $\omega$ implies one independent complex analysis of equation [3.57] if the FM is used. The reason why the exact PEM is not suggested in topology optimization lies in the fact that the computing cost is prohibitive in practice for the multiple excitation frequencies.

### 3.4.4. *Comparative tests of stationary random analysis methods*

Two examples are dealt with by means of the conventional PEM, the combined method of PEM and MAM as well as the exact PEM. In all numerical tests of this chapter, Young's modulus, Poisson ratio and density of the solid material are set to be 200 GPa, 0.3 and 7,800 kg/m$^3$, respectively. The same Rayleigh damping is also adopted with $\alpha = 10^{-2}$ and $\beta = 10^{-5}$. To simplify the discussion, the combined method of PEM and MAM is termed improved PEM.

*Structure 1: 2D cantilever beam*

The structure is a rectangular domain of size 0.8 m × 0.4 m × 0.001 m. It is clamped at the left side, as shown in Figure 3.16. Here, the domain is meshed into 40 × 20 plane stress elements with 1,722 DOFs in all. The white-noise force excitation of PSD value 2,500 N$^2$/(rad/s) is applied at the middle node of the right edge.

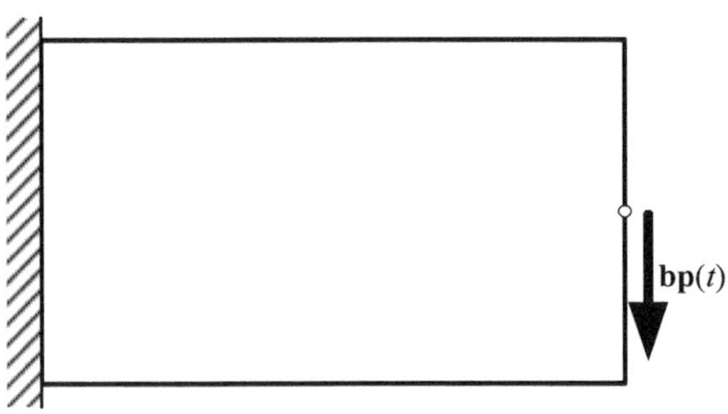


**Figure 3.16.** *2D cantilever beam (800 plane elements, 1,722 DOFs)*

*Structure 2: 3D beam*

The beam structure has a size of 0.5 m × 0.3 m × 0.2 m. It is simply supported at four corners, as shown in Figure 3.17. The domain is meshed into 50 × 30 × 20 solid elements with 99,603 DOFs in all. The white-noise force excitation of PSD value $10^4$ $N^2/(rad/s)$ is applied at the middle bottom point of the other side.

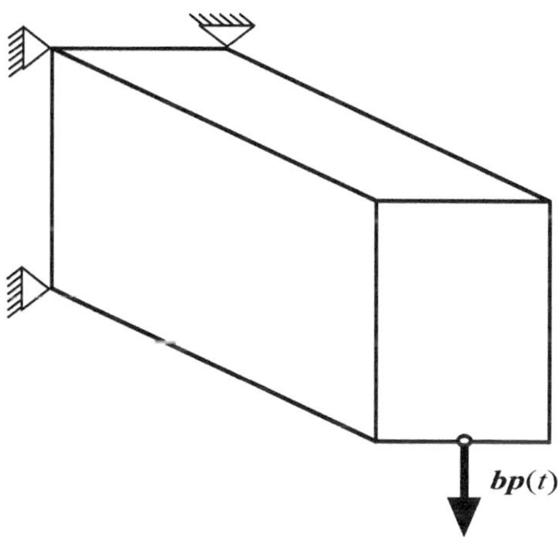


**Figure 3.17.** *3D beam (30,000 solid elements, 99,603 DOFs)*

Here, the first $l = 30$ modes are employed in both examples. Notice that structure 1 represents the common problem studied previously with a discretization of roughly thousands of DOFs. Structure 2 is adopted to illustrate the large-scale problems of huge numbers of DOFs that are rarely studied in dynamic topology optimization.

According to equation [3.54], the PSD value of random displacement at the loading point along the force direction is calculated as:

$$S_{u_s}(\omega) = \sum_{q=1}^{Q} (\mathbf{g}_q)_s^* (\mathbf{g}_q)_s^{\mathrm{T}} = \sum_{q=1}^{Q} \left| (\mathbf{g}_q)_s \right|^2 \quad [3.58]$$

where $s$ denotes the concerned DOF number. The solution of equation [3.58] will be obtained by the exact FM, the improved PEM as well as the conventional method. Suppose

$$\Omega(\omega)=\log\left(S_{u_s}(\omega)\right) \qquad [3.59]$$

Results are then compared in the above form of logarithm in Figures 3.18 and 3.19. Notice that the excitation frequency ($\omega = 2\pi f$) in each problem is always covered by the first $l$ eigenfrequencies by default.

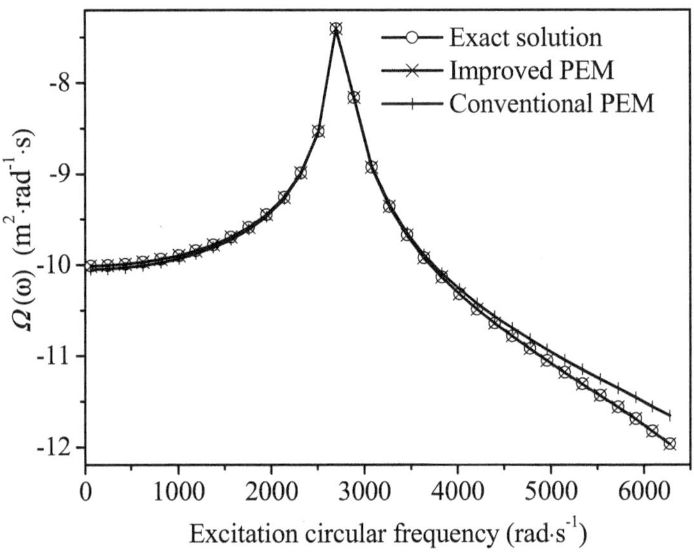


**Figure 3.18.** *PSD curves obtained using three different methods for structure 1*

In both examples, it is shown that the PSD curves using the improved PEM are very close to the exact solution. Instead, the conventional PEM is acceptable only when the excitation frequency is low. The error is relatively small near the peaks of PSD curves, where the excitation frequency attains the resonant eigenfrequency. In

Figures 3.18 and 3.19, it is also seen that the accuracy of the conventional PEM decreases with the dramatic increase in the DOF number, while the improved PEM maintains its accuracy. Hence, the improved PEM outperforms the conventional PEM in terms of accuracy for large-scale problems.

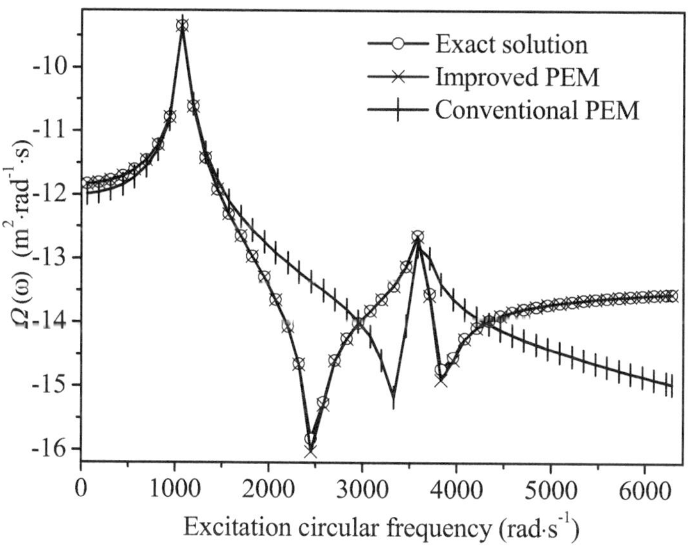


**Figure 3.19.** *PSD curves obtained using three different methods for structure 2*

## 3.5. Topology optimization under stationary random force excitation

### 3.5.1. *Topology optimization formulation*

The root mean square (RMS) of random response can be used to represent the vibration level in practice. Here, the RMS of random displacement response of the concerned $r$th DOF is considered as the objective function.

$$R_{u_r} = \sqrt{\int_{\omega_A}^{\omega_B} S_{u_r}(\omega)d\omega} \qquad [3.60]$$

$[\omega_A, \omega_B]$ refers to the frequency interval of random excitation. Therefore, dynamic topology optimization is stated as:

$$\begin{array}{ll} \text{find} & \boldsymbol{\eta} = \{\eta_i\} \quad i = 1,2,\ldots,n_e \\ \min & R_{u_s} \\ \text{s.t} & V \le V_U \\ & 0 < \eta_L \le \eta_i \le 1 \end{array} \qquad [3.61]$$

### 3.5.2. *Sensitivity analysis*

The sensitivity of RMS with respect to pseudo-density variable $\eta_i$ is presented below. The differentiation of equation [3.60] can be written as:

$$\frac{\partial R_{u_s}}{\partial \eta_i} = \frac{1}{2R_{u_s}} \int_{\omega_A}^{\omega_B} \frac{\partial S_{u_s}(\omega)}{\partial \eta_i} d\omega \qquad [3.62]$$

According to equation [3.58], the following equation holds:

$$\frac{\partial S_{u_s}(\omega)}{\partial \eta_i} = \sum_{q=1}^{Q} 2\left|(\mathbf{g}_q)_s\right| \frac{\partial \left|(\mathbf{g}_q)_s\right|}{\partial \eta_i} \qquad [3.63]$$

Clearly, the sensitivity of displacement amplitude of pseudo-harmonic response is the basic calculation for equation [3.62]. With the implementation of the MDM, the sensitivity of pseudo-harmonic displacement can be obtained by directly differentiating equation [3.53].

$$\frac{\partial \mathbf{g}_q}{\partial \eta_i} = \sum_{k=1}^{l} \left( \frac{\partial \boldsymbol{\varphi}_k}{\partial \eta_i} H_k \boldsymbol{\varphi}_k^{\mathrm{T}} + \boldsymbol{\varphi}_k \frac{\partial H_k}{\partial \eta_i} \boldsymbol{\varphi}_k^{\mathrm{T}} + \boldsymbol{\varphi}_k H_k \frac{\partial \boldsymbol{\varphi}_k^{\mathrm{T}}}{\partial \eta_i} \right) \mathbf{b} \boldsymbol{\gamma}_q e^{\iota \omega t} \qquad [3.64]$$

With the implementation of the MAM, the sensitivity of pseudo-harmonic displacement corresponds to

$$\frac{\partial \mathbf{g}_q}{\partial \eta_i} = \frac{\partial\left(\mathbf{K}^{-1}\mathbf{b}\boldsymbol{\gamma}_q\right)}{\partial \eta_i} e^{\iota\omega t} + \sum_{k=1}^{l}\left(\frac{\partial \boldsymbol{\varphi}_k}{\partial \eta_i} Z_k \boldsymbol{\varphi}_k^{\mathrm{T}} + \boldsymbol{\varphi}_k \frac{\partial Z_k}{\partial \eta_i} \boldsymbol{\varphi}_k^{\mathrm{T}} + \boldsymbol{\varphi}_k Z_k \frac{\partial \boldsymbol{\varphi}_k^{\mathrm{T}}}{\partial \eta_i}\right)\mathbf{b}\boldsymbol{\gamma}_q e^{\iota\omega t} \quad [3.65]$$

where $Z_k$ is defined in equation [3.25].

### 3.5.3. *Numerical examples*

In this section, topology optimization problems under stationary random force excitation are addressed. The conventional PEM and the improved PEM are used to solve large-scale problems. The GCMMA algorithm is used as the optimizer.

#### *Structure 1*

The problem is illustrated in Figure 3.16. Suppose the RMS of vertical displacement at the loading position is minimized. The volume fraction of solid material is constrained to be less than 50% of the design domain. Initial values of all pseudo-densities are set to be 0.5. Two frequency intervals of random force excitation are considered with $f$ = [0, 100] Hz and [0, 500] Hz. Here, $l$ = 30 modes are employed with the first eigenfrequency being 152 Hz and the 30th eigenfrequency being 4,541 Hz for the initial structure.

Nearly the same configurations are obtained by the conventional PEM and the improved PEM, as shown in Figures 3.20 and 3.21. Exact solutions of PSD curves are shown in Figure 3.22 for the optimized configurations related to Figure 3.21. It can be seen that the PSD within the prescribed optimization frequency intervals decreases obviously for the minimization of the RMS of random displacement response.

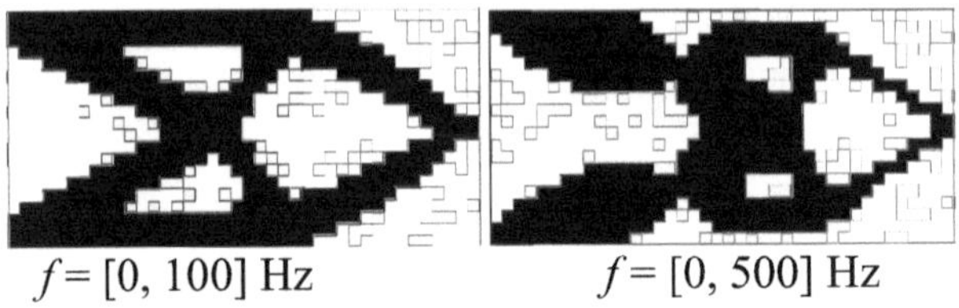


**Figure 3.20.** *Optimized configurations of structure 1 by the conventional PEM* (1,722 DOFs)

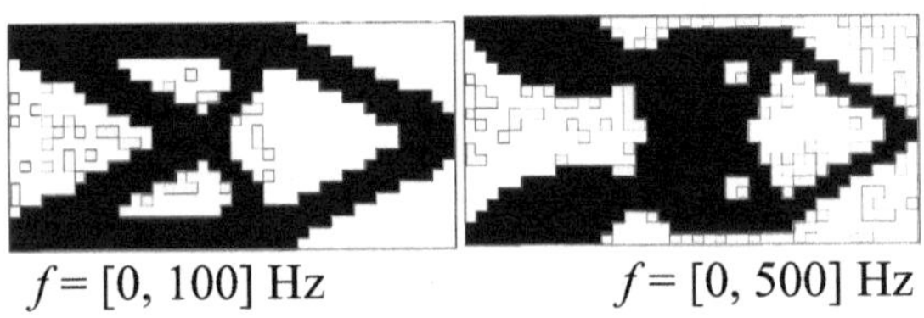


**Figure 3.21.** *Optimized configurations of structure 1 by the improved PEM (1,722 DOFs)*

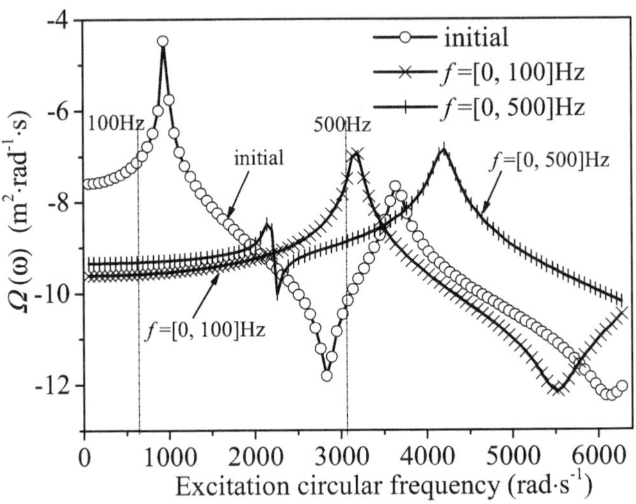


**Figure 3.22.** *Exact solutions of PSD curves for optimized configurations in Figure 3.21*

Other problems with the similar order of DOFs can also be found in [LIN 11, RON 08, RON 00, ZHA 10a]. Therefore, both the conventional PEM and the improved PEM are effective when the number of DOFs is small. Now, structure 1 will be further optimized with refined meshes of 120 × 60 elements and large numbers of DOFs, while $l$ = 30 modes are still employed. Figure 3.23 indicates that unclear configurations, especially near the loading point, are obtained by the conventional PEM with a poor convergence. In contrast, the improved PEM works well and produces clear configurations as shown in Figure 3.24.

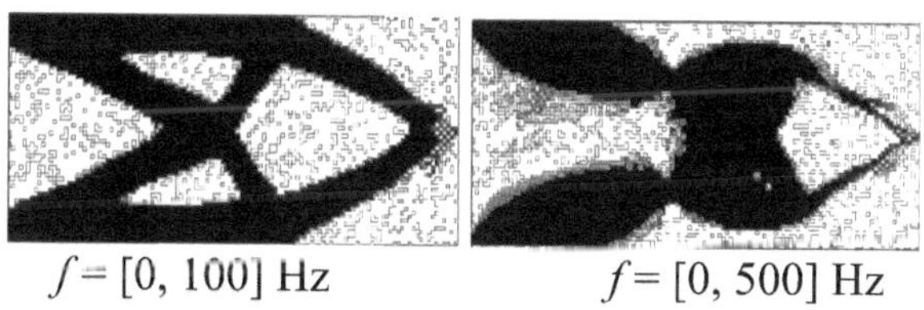


**Figure 3.23.** *Optimized configurations of structure 1 by the conventional PEM (14,762 DOFs)*

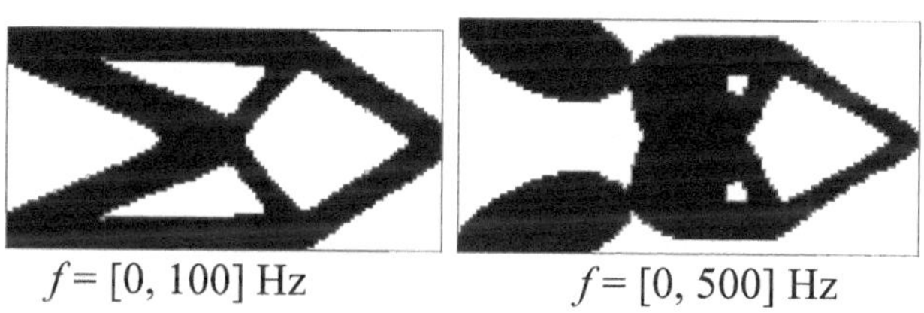


**Figure 3.24.** *Optimized configurations of structure 1 by the improved PEM (14,762 DOFs)*

### *Structure 2*

To further verify the effects of the improved PEM, consider now the 3D problem illustrated in Figure 3.17. Suppose the RMS of random displacement response at the loading position along the force direction is minimized. The volume fraction of solid material is constrained to be less than 20%. Initial values of all pseudo-densities are set to be 0.2. Two frequency intervals are considered with $f$ = [0, 100] Hz and [0, 500] Hz. $l$ = 30 modes are still employed with the first eigenfrequency being 30 Hz and the 30th eigenfrequency being 2,476 Hz for the initial structure. Configurations optimized by the conventional PEM and the improved PEM are shown in Figures 3.25 and 3.26, respectively.

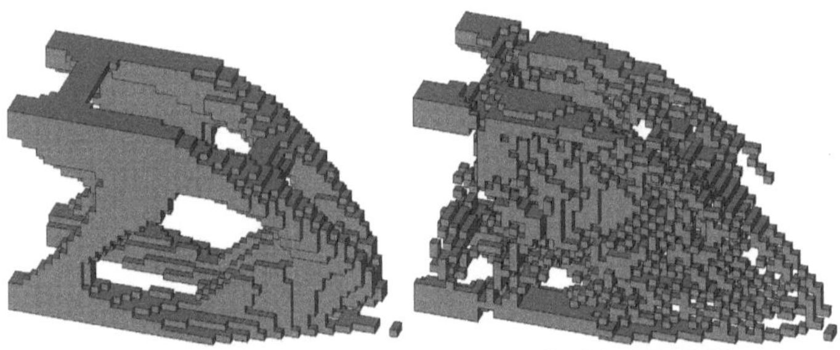

$f = [0, 100]$ Hz $f = [0, 500]$ Hz

**Figure 3.25.** *Optimized configurations of structure 2 by the conventional PEM* (99,603 DOFs)

Likewise, it is difficult for the conventional PEM to achieve convergence, especially with the expanding interval of excitation frequency. Comparatively, the improved PEM is very efficient in both frequency intervals. Optimized configurations and exact solutions of PSD curves are shown in Figures 3.26 and 3.27, respectively. It is shown that PSD curves of optimized structures globally decrease within the prescribed frequency intervals.

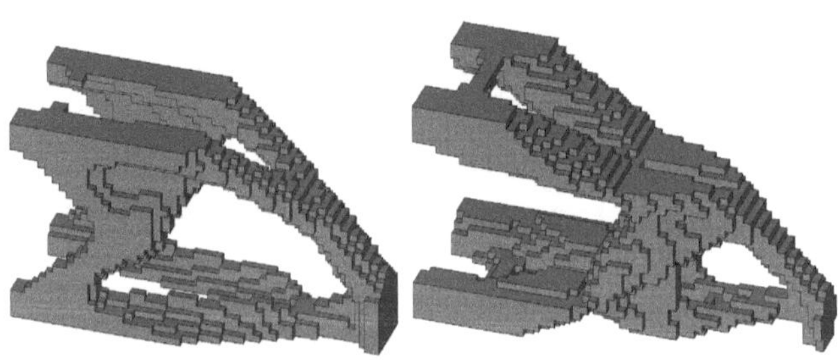

$f = [0, 100]$ Hz $f = [0, 500]$ Hz

**Figure 3.26.** *Optimized configurations of structure 2 by the improved PEM (99,603 DOFs)*

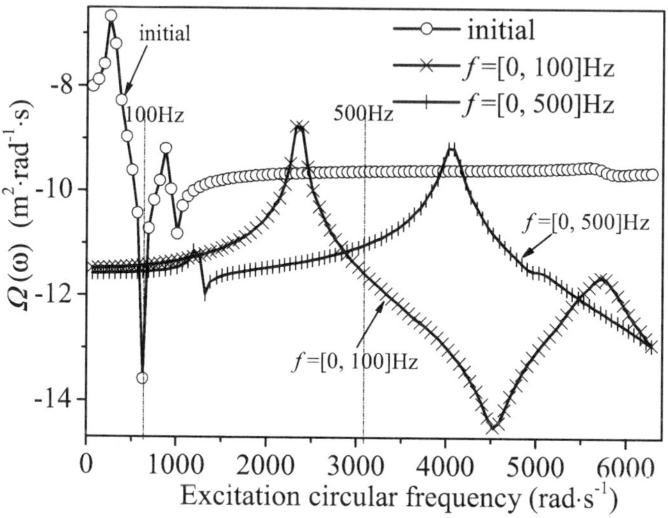


**Figure 3.27.** *Exact solutions of PSD curves for optimized configurations shown in Figure 3.26*

### *Structure 3*

A bracket structure is illustrated in Figure 3.28. It is clamped by four bolts and loaded by a white-noise force excitation at the middle node of the top edge with PSD value of $10^4$ $N^2$/(rad/s). Suppose the RMS of displacement at the loading position along the force direction is minimized. Notice that materials around the loading point and four bolt holes are set as non-designable solids. The volume fraction of solid material in the design domain is constrained to be less than 10%. Initial values of all pseudo-densities of elements in the design domain are set to be 0.1. Two frequency intervals are considered with $f$ = [0, 400] Hz and [0,3000] Hz. Here, $l$ = 20 modes are employed with the first eigenfrequency being 356 Hz and the 20th eigenfrequency being 5,537 Hz for the initial structure. Within the above frequency intervals, optimized configurations shown in Figures 3.29 and 3.30 are obtained by the conventional PEM and the improved PEM. As indicated in Figure 3.31, PSD curves related to optimized structures given in Figure 3.30 globally decrease within the prescribed frequency intervals.

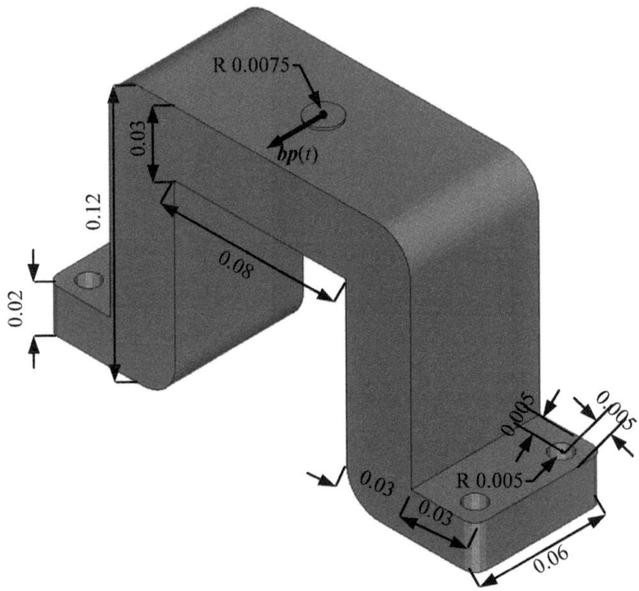


**Figure 3.28.** *Bracket structure (42,300 solid elements, 145,104 DOFs)*

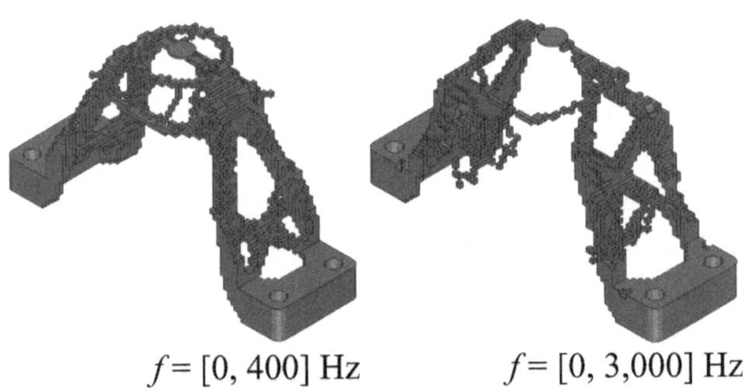


**Figure 3.29.** *Optimized configurations of structure 3 by the conventional PEM* (145,104 DOFs)

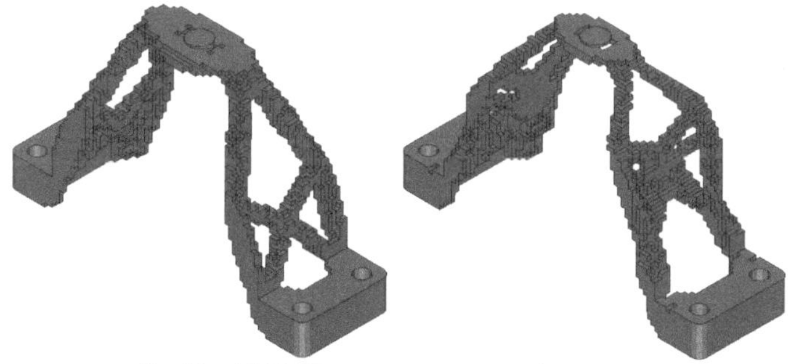


**Figure 3.30.** *Optimized configurations of structure 3 by the improved PEM (145,104 DOFs)*

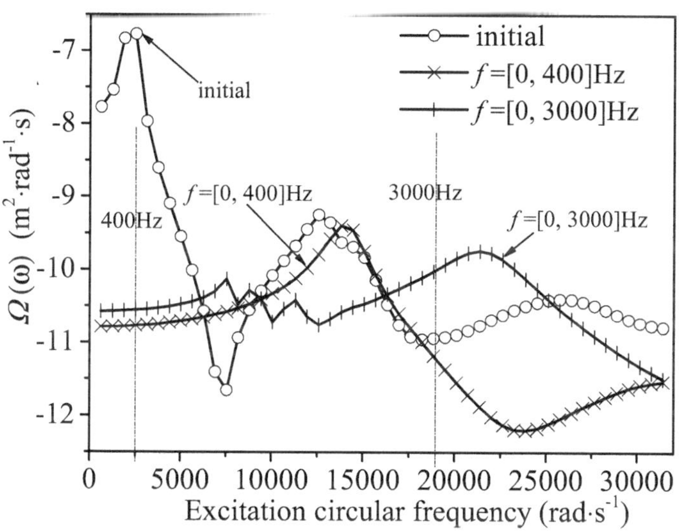


**Figure 3.31.** *Exact solutions of PSD curves for the optimized configurations given in Figure 3.30*

From the above examples, it may be concluded that the conventional PEM is only limited to small-scale problems. For large-scale problems the convergence is very poor, especially at high excitation frequency intervals. It is almost impossible to identify clear configurations from obtained results. In contrast, the improved PEM is efficient in all cases for dynamic topology optimization.

## 3.6. Conclusions

In topology optimization related to harmonic responses, the MDM was generally adopted as the dynamic analysis method. However, unsatisfactory convergences become inevitable due to the low computing accuracy of harmonic responses when large-scale problems are concerned in practice. In this chapter, effects of the MDM on topology optimization results are investigated, and the reasons for the low accuracy of structural response are discussed to highlight the errors caused by the truncation modes and by the augmented gap between excitation frequency and eigenfrequencies of optimized structures. It is shown that the MDM would become critical with the increase in DOFs. Therefore, both the MAM and the FM are proposed to improve the computing accuracy and the convergence of large-scale topology optimization problems. Theoretical analysis and numerical tests demonstrate that the FM outperforms the MDM and MAM in the aspect of analysis accuracy and efficiency in the case of harmonic excitation with one specific frequency. For the optimization problems under harmonic excitations with multiple frequencies, the MAM is suggested because of its compromise between computing accuracy and efficiency.

In topology optimization related to random responses, the CQC was generally adopted as the dynamic analysis method. However, prohibitive computing cost and unsatisfactory convergences became inevitable when large-scale problems were concerned in practice. The introduction of the conventional PEM can greatly improve efficiency, while unsatisfactory convergence is still a great difficulty due to the low accuracy of the CQC and PEM. Therefore, the improved PEM is introduced in the current work. The advantage is twofold: the embedded PEM can improve the computing efficiency in the optimization, while the computing accuracy of random response is guaranteed by means of the embedded MAM. Finally, theoretical analysis and numerical tests demonstrate that the proposed optimization procedure outperforms the existing methods in terms of efficiency and convergence, and opens its great practicability in dynamic topology optimization of large-scale structures subjected to stationary random excitations.

# 4

# Thermo-Elastic Problems

## 4.1. Introduction

Thermo-elastic topology optimization is complicated because it belongs to a kind of design-dependent problem [ZHA 14b] with the thermal stress load changing along with the spatial distribution of solid material phases. Generally, the aim is to achieve one such design that produces a optimized structure that is stiff enough to support the mechanical load and compliant enough in proper areas to release the thermal stress.

Most works on thermo-elastic topology optimization have been limited to the case of one single material phase. For example, Rodrigues and Fernandes [ROD 95] adopted the homogenization method to formulate the thermal stress load for the mean compliance minimization. Li *et al.* [LI 99] used the ESO method with element thickness to be design variables. An adjoint design sensitivity analysis method [CHO 05] was developed for the topology optimization of weakly coupled thermo-elastic problems. Structural rigidity optimization with an initial design-dependent thermo-elastic stress field was also presented [DES 13]. Deng *et al.* [DEN 13] optimized the microstructure of homogeneous porous material and macrostructure topology. Pedersen and Pedersen [PED 12] found that minimization of the maximum von-Mises stress can be achieved through accomplishing a uniform energy density recursively. Recent results from Zhang *et al.* [ZHA 14a] indicated that the elastic strain energy minimization and mean compliance minimization led to different configurations if thermal loads exist. The elastic strain energy minimization particularly favors stress reduction.

Meanwhile, multiple materials were taken into account and the concept of thermal stress coefficient (TSC) defined as the product between Young's modulus and coefficient of thermal expansion was introduced [GAO 10]. The TSC was adopted later in the thermo-elastic topology optimization of stress-constrained problems [DEA 13] and dynamic compliance minimization [YAN 14].

The outline of this chapter is as follows. First, the concept of TSC is introduced to model the thermal stress load. Second, suppose that only one solid material is used. Thermo-elastic topology optimization is formulated as the minimization of the structural compliance subjected to the volume constraint. Numerical tests are performed to illustrate the formulation and comparison is made between the SIMP and RAMP models. Third, a general formulation is proposed for thermo-elastic topology optimization with multiple materials and mass constraint. Finally, the mean compliance and strain energy are compared for the definition of the objective function.

## 4.2. Thermo-elastic analysis

The particularity of the thermo-elastic topology optimization is that the thermal stress load relies upon both the Young's modulus and the coefficient of thermal expansion. The inherent issue is therefore how to establish the formulation in the case of the thermal stress load.

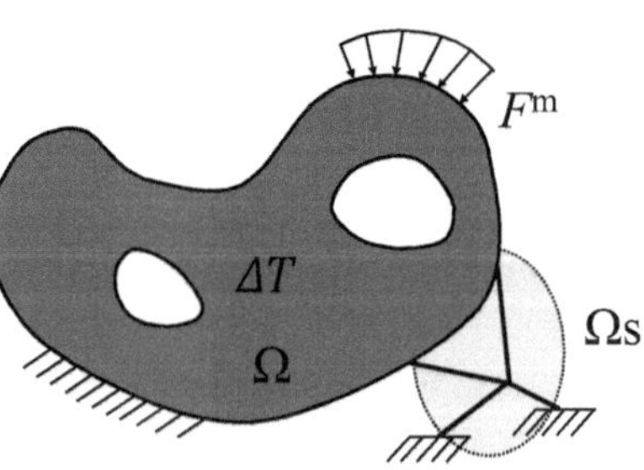


**Figure 4.1.** *Elastic body under thermal and mechanical loads*

First, it is necessary to recall some basic concepts. For a thermo-elastic structure undergoing thermal and mechanical loads, the stress-strain relation reads

$$\begin{aligned} \boldsymbol{\sigma}_i &= \mathbf{D}\boldsymbol{\varepsilon}_i^{\mathrm{m}} \\ \boldsymbol{\varepsilon}_i^{\mathrm{m}} &= \boldsymbol{\varepsilon}_i - \boldsymbol{\varepsilon}_i^{\mathrm{th}} = \partial \mathbf{U}_i - \boldsymbol{\varepsilon}_i^{\mathrm{th}} \end{aligned} \qquad [4.1]$$

where $\mathbf{D}$ is the Hooke matrix, $\mathbf{U}_i$ is the displacement that is often calculated by the FEM and $\boldsymbol{\varepsilon}_i$ is the strain. Notice that superscripts m and th denote mechanical and thermal parts, respectively. As shown in Figure 4.1, designs can be made to improve the structural rigidity or to reduce the thermal stress by optimizing the material layout either over the main structural domain Ω or the elastic support Ωs.

In this chapter, the finite element equilibrium equation corresponds to

$$\mathbf{KU} = \mathbf{F} = \mathbf{F}^{\mathrm{m}} + \mathbf{F}^{\mathrm{th}} \qquad [4.2]$$

Here, $\mathbf{K}$ is the global stiffness matrix of the structure. $\mathbf{U}$ is the nodal displacement vector. $\mathbf{F}^{\mathrm{m}}$ and $\mathbf{F}^{\mathrm{th}}$ are nodal force vectors related to mechanical and thermal loads, respectively.

The nodal thermal stress force vector of the *i*th element can be written as

$$\mathbf{F}_i^{\mathrm{th}} = \int_{\Omega_i} \mathbf{B}_i^{\mathrm{T}} \mathbf{D}_i \boldsymbol{\varepsilon}_i^{\mathrm{th}} d\Omega \qquad [4.3]$$

By definition, $\mathbf{B}_i$ and $\mathbf{D}_i$ are element strain-displacement matrix and elasticity matrix, respectively. The former consists of derivatives of element shape functions that are independent of topology design variables. In this work, isotropic linear elastic materials are used, and the elasticity matrix $\mathbf{D}$ can be written as:

$$\mathbf{D}_i = \frac{E_i(1-\mu_i)}{(1+\mu_i)(1-2\mu_i)} \begin{bmatrix} \begin{bmatrix} 1 & \frac{\mu_i}{1-\mu_i} & \frac{\mu_i}{1-\mu_i} \\ \frac{\mu_i}{1-\mu_i} & 1 & \frac{\mu_i}{1-\mu_i} \\ \frac{\mu_i}{1-\mu_i} & \frac{\mu_i}{1-\mu_i} & 1 \end{bmatrix} & \mathbf{0}_{3\times 3} \\ \mathbf{0}_{3\times 3} & \frac{1-2\mu_i}{2(1-\mu_i)} \begin{bmatrix} 1 & 0 & 0 \\ 0 & 1 & 0 \\ 0 & 0 & 1 \end{bmatrix} \end{bmatrix} \quad [4.4]$$

Suppose that the coefficient of thermal expansion is temperature-independent. Thus, the thermal strain vector $\boldsymbol{\varepsilon}_i^{\text{th}}$ is written as:

$$\boldsymbol{\varepsilon}_i^{\text{th}} = \boldsymbol{\alpha}_i \Delta T_i \quad [4.5]$$

where $\Delta T_i$ denotes the temperature rise of the $i$th element. The vector of the coefficient of thermal expansion, $\boldsymbol{\alpha}_i$, can be written as:

$$\begin{aligned} \boldsymbol{\alpha}_i &= \alpha_i \boldsymbol{\phi} \\ \boldsymbol{\phi} &= \begin{bmatrix} 1 & 1 & 1 & 0 & 0 & 0 \end{bmatrix}^{\mathrm{T}} \end{aligned} \quad [4.6]$$

where $\alpha_i$ is the coefficient of thermal expansion. The substitution of $\boldsymbol{\varepsilon}_i^{\text{th}}$ into equation [4.3] then produces

$$\mathbf{F}_i^{\text{th}} = \int_{\Omega_i} \mathbf{B}_i^{\mathrm{T}} \mathbf{D}_i \boldsymbol{\alpha}_i \Delta T_i d\Omega \quad [4.7]$$

Then, the vector of the thermal stress coefficient (TSC) is defined as:

$$\boldsymbol{\beta}_i = \mathbf{D}_i \boldsymbol{\alpha}_i \quad [4.8]$$

The substitution of equations [4.4] and [4.6] into equation [4.8] yields

$$\begin{aligned} \boldsymbol{\beta}_i &= \beta_i \boldsymbol{\phi} \\ \beta_i &= \frac{E_i \alpha_i}{1-2\mu_i} \end{aligned} \quad [4.9]$$

Thus, TSC can be treated as an inherent material property. $\mathbf{F}_i^{\text{th}}$ is then rewritten as:

$$\begin{aligned} \mathbf{F}_i^{\text{th}} &= \beta \overline{\mathbf{F}}_i^{\text{th}} \\ \overline{\mathbf{F}}_i^{\text{th}} &= \int_{\Omega_i} \mathbf{B}_i^{\text{T}} \boldsymbol{\phi} \Delta T_i d\Omega \end{aligned} \qquad [4.10]$$

## 4.3. Thermo-elastic topology optimization with single material

In this section, the basic thermo-elastic topology optimization is presented to minimize the structural compliance subjected to the volume constraint. Here, single solid material and void are considered. The models and conclusions can be extended into design problem with multiple materials studied later.

### 4.3.1. *Topology optimization formulation*

The topology optimization problem involving thermo-elastic stress load can be expressed as a compliance minimization subjected to the volume constraint.

$$\begin{aligned} &\text{find} && \boldsymbol{\eta} = \{\eta_i\} \quad i = 1,2,\ldots,n_{\text{e}} \\ &\min && C = \mathbf{F}^{\text{T}}\mathbf{U} \\ &\text{s.t} && \mathbf{F} = \mathbf{K}\mathbf{U} \\ & && \mathbf{Q} = \mathbf{H}\mathbf{T} \\ & && V_{\text{C}} = \sum_i V_i \eta_i \le V_{\text{U}} = vf_{\text{U}} \cdot V \\ & && 0 < \eta_{\text{L}} \le \eta_i \le 1 \end{aligned} \qquad [4.11]$$

where **Q**, **T** and **U** are the global thermal flux vector, nodal temperature vector and nodal displacement vector. **K** is the global stiffness matrix and **H** is the global heat conductivity matrix. $V_{\text{U}}$ denotes the upper bound of the volume constraint of the candidate material $V_{\text{C}}$. The upper bound of the volume fraction, $vf_{\text{U}}$, is defined as the ratio of $V_U$ to the total volume of all designable elements $V$ with $vf_U = V_U/V$. Obviously, $0 < vf_U < 1$ for a meaningful volume constraint.

### 4.3.2. *Sensitivity analysis*

As detailed in Chapter 2, the SIMP interpolation model of exponential form might lead to a mismatch between force and stiffness for low-density elements with very small values of $\eta_i$ when design-dependent body forces such as self-weight and centrifugal forces are present. Here, the RAMP model is adopted with

$$E_i = \frac{\eta_i}{1+q_E(1-\eta_i)}E_0$$
$$\beta_i = \frac{\eta_i}{1+q_\beta(1-\eta_i)}\beta_0 \qquad [4.12]$$
$$\kappa_i = \frac{\eta_i}{1+q_\kappa(1-\eta_i)}\kappa_0$$

Here, interpolations of the Young's modulus ($E$), thermal stress coefficient ($\beta$) and heat conductivity coefficient ($\kappa$) are made simultaneously to compute element stiffness matrix and thermal stress load. The subscript 0 denotes the solid material. In the case of a single material, the Poisson's ratio is supposed to be constant and $\beta_0 = E_0\alpha_0/(1\text{-}2\mu_0)$.

When gradient-based optimizers are used to solve topology optimization problems, it is necessary to carry out a sensitivity analysis with respect to pseudo-density variables. The sensitivity of the structural compliance corresponds to

$$\frac{\partial C}{\partial \eta_i} = 2\mathbf{U}^{\mathrm{T}}\left(\frac{\partial \mathbf{F}^{\mathrm{m}}}{\partial \eta_i}+\frac{\partial \mathbf{F}^{\mathrm{th}}}{\partial \eta_i}\right)-\mathbf{U}^{\mathrm{T}}\frac{\partial \mathbf{K}}{\partial \eta_i}\mathbf{U} = 2\mathbf{U}_i^{\mathrm{T}}\frac{\partial \mathbf{F}_i^{\mathrm{th}}}{\partial \eta_i}-\mathbf{U}_i^{\mathrm{T}}\frac{\partial \mathbf{K}_i}{\partial \eta_i}\mathbf{U}_i \qquad [4.13]$$

Term $\partial\mathbf{F}^{\mathrm{m}}/\partial\eta_i$ is often equal to zero when $\mathbf{F}^{\mathrm{m}}$ is design-independent except for inertial forces, such as gravity and centrifugal loads.

Evidently, $\partial \mathbf{K}_i/\partial \eta_i$ and $\partial \mathbf{F}_i^{\text{th}}/\partial \eta_i$ can easily be derived at the element level.

For the $i$th element, the stiffness matrix is expressed as

$$\mathbf{K}_i = \int_{\Omega_i} \mathbf{B}_i^{\mathrm{T}} \mathbf{D}_i \mathbf{B}_i \, d\Omega \tag{4.14}$$

Mathematically, $\partial \mathbf{K}_i/\partial \eta_i$ can be written as

$$\begin{aligned} \frac{\partial \mathbf{K}_i}{\partial \eta_i} &= \int_{\Omega} \mathbf{B}_i^{\mathrm{T}} \frac{\partial \mathbf{D}_i}{\partial \eta_i} \mathbf{B}_i \, d\Omega = \frac{1+q_E}{\left(1+q_E\left(1-\eta_i\right)\right)^2} \mathbf{K}_{i,0} \\ \mathbf{K}_{i,0} &= \int_{\Omega_i} \mathbf{B}_i^{\mathrm{T}} \mathbf{D}_0 \mathbf{B}_i \, d\Omega \end{aligned} \tag{4.15}$$

where $\mathbf{D}_0$ is the elasticity matrix related to the solid material. As to term $\partial \mathbf{F}_i^{\text{th}}/\partial \eta_i$, it is derived from equation [4.10].

$$\frac{\partial \mathbf{F}_i^{\text{th}}}{\partial \eta_i} = \frac{\partial \beta_i}{\partial \eta_i} \overline{\mathbf{F}}_i^{\text{th}} + \beta_i \frac{\partial \overline{\mathbf{F}}_i^{\text{th}}}{\partial \eta_i} = \frac{\partial \beta_i}{\partial \eta_i} \overline{\mathbf{F}}_i^{\text{th}} + \beta_i \int_{\Omega_i} \mathbf{B}_i^{\mathrm{T}} \boldsymbol{\phi} \frac{\partial \Delta T_i}{\partial \eta_i} d\Omega \tag{4.16}$$

Theoretically, two cases exist in the calculation of $\partial \Delta T_i/\partial \eta_i$: $\partial \Delta T_i/\partial \eta_i \equiv 0$ for a design-independent temperature field whatever the material layout. $\partial \Delta T_i/\partial \eta_i \neq 0$ for a design-dependent temperature field. In our implementation, a constant temperature field (zero-order approximation) is assumed in sensitivity analysis, while the temperature field is updated on the finite element analysis level of steady-state heat conduction. In this way, the efficiency of the design procedure can be largely improved without deterioration of the design accuracy. Thus, the sensitivity of the thermal stress force vector of the $i$th element can be approximately simplified as:

$$\frac{\partial \mathbf{F}_i^{\text{th}}}{\partial \eta_i} = \frac{\partial \beta_i}{\partial \eta_i} \overline{\mathbf{F}}_i^{\text{th}} \tag{4.17}$$

### 4.3.3. *Numerical examples*

#### 4.3.3.1. *Constant temperature rise*

*Numerical comparisons between SIMP and RAMP models*

A biclamped structure with two non-designable domains (the dark areas) is shown in Figure 4.2. It was studied previously by Rodrigues and Fernandes [ROD 95]. A mesh of 60 × 40 elements is used here. Assume that Young's modulus of the solid material is 210 GPa, the coefficient of thermal expansion is $1.1\times10^{-5}$/°C and the Poisson's ratio is 0.3. The applied force has a value of 10 kN and a uniform temperature rise is 1 K over the domain.

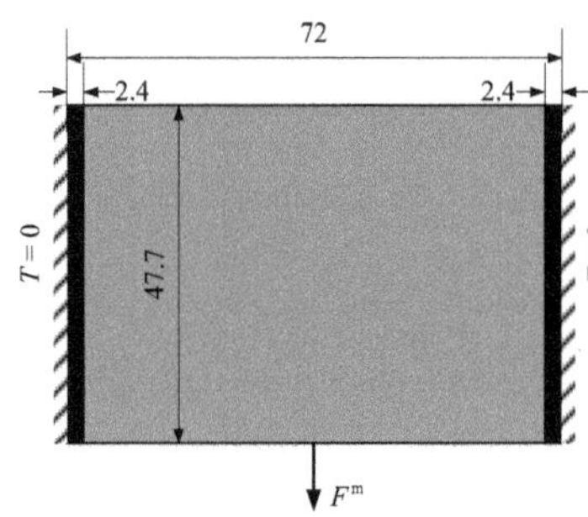


**Figure 4.2.** *Structure 1: biclamped rectangular domain (unit: cm)*

The upper bound of the volume fraction is $vf_U = 0.4$. The SIMP model is applied first. With one solid material and void, SIMP models of the Young's modulus and TSC correspond to $E_i = \eta_i^{p_E} E_0$ and $\beta_i = \eta_i^{p_\beta} \beta_0$. As usual, the penalty factor $p_E = 3$ is used for Young's modulus here. For different values of penalty factor for TSC ($p_\beta$), iteration curves of the normalized compliance and optimized configurations are shown in Figure 4.3. It is found that the iteration history is unstable and small oscillations exist at the end of iteration when $p_\beta = 1$. Meanwhile, the parasitic effect of the material layout is quite evident. For $p_\beta > 1$, although the iteration history becomes stable, the unexpected gray areas exist for $p_\beta > 2$ and the gray area increases. A clear configuration is obtained only for $p_\beta = 2$. Unfortunately,

Figure 4.4 indicated that the SIMP model is unable to yield a clear configuration if the uniform temperature rise is 3K with $p_\beta = 2$ and $p_E = 3$.

Alternatively, the RAMP model is now applied for the same problem. Iteration histories and the optimization results for $\Delta T = 1$ K and $\Delta T = 3$ K are shown in Figures 4.5 and 4.6, respectively. With $q_E = 8$ and $q_\beta = 0$, the RAMP model leads to stable iterations and clear configurations in both cases. Accordingly, the RAMP model seems to be more robust than the SIMP model for this kind of problem.

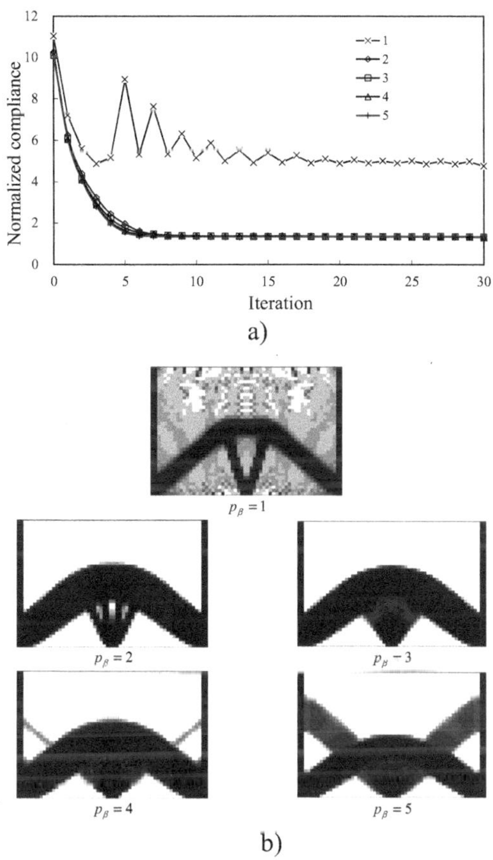


**Figure 4.3.** *Optimization design with SIMP model ($\Delta T = 1$ K, $p_E = 3$): a) iteration histories of the normalized compliance with different $p_\beta$; b) structural configuration at last iteration*

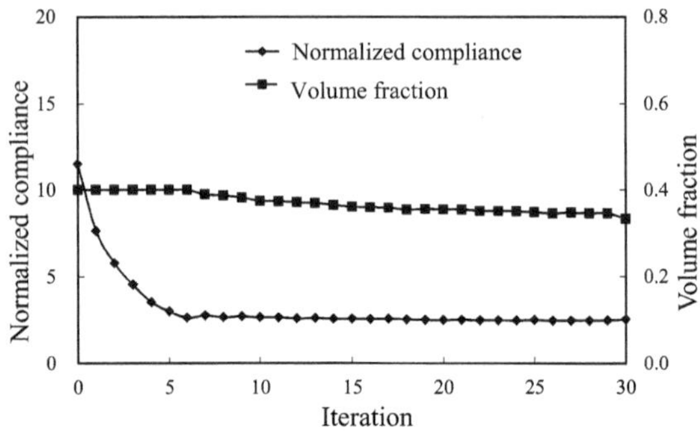


a)

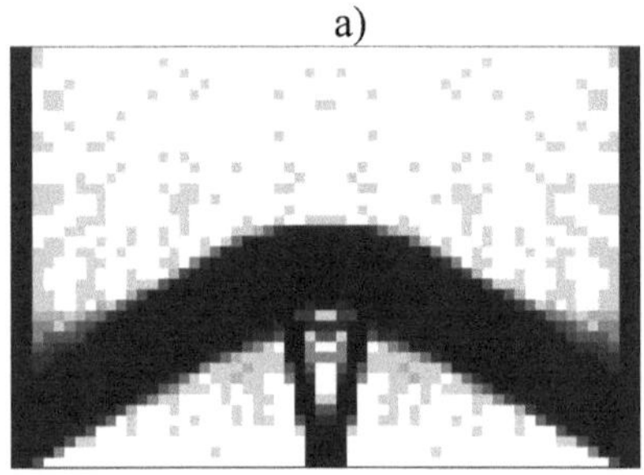

b)

**Figure 4.4.** *Optimization design with SIMP model ($\Delta T$ = 3 K $p_E$ = 3, $p_\beta$ = 2): a) iteration histories of the normalized compliance and the volume fraction; b) structural configuration at last iteration*

It should be remarked that at $\Delta T$ = 1 K, the volume fraction always attains the prescribed upper bound for both SIMP and RAMP models. At $\Delta T$ = 3 K, Figures 4.4(a) and 4.6(a) indicate that the volume fraction constraint behaves differently for each model. The RAMP produces an optimization structure slightly stiffer than using the SIMP model.

In fact, from the work of Stolpe and Svanberg [STO 01], it is known that the RAMP is always first-order infinitesimal with non-zero slope when the element pseudo-density tends to zero. In contrast, the SIMP has a zero slope when the element pseudo-density tends to zero if the penalty factor is larger than 1. This difference means that both the element stiffness and the thermal load related to the SIMP model are insensitive to the zero value of pseudo-density.

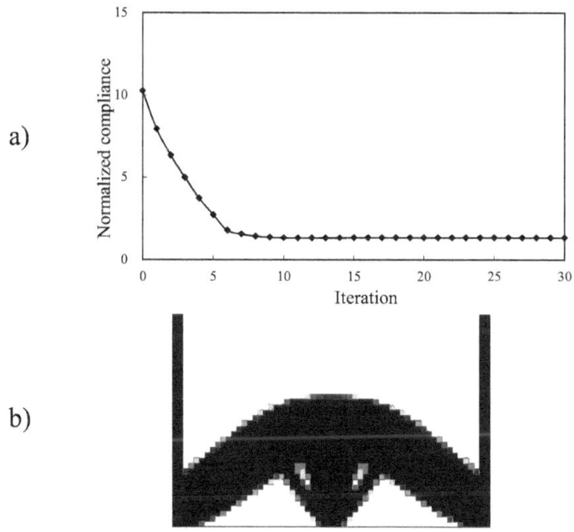


**Figure 4.5.** *Optimization design with RAMP model ($\Delta T$ = 1 K, $q_E$ = 8, $q_\beta$ = 0): a) iteration history of the normalized compliance; b) structural configuration at last iteration*

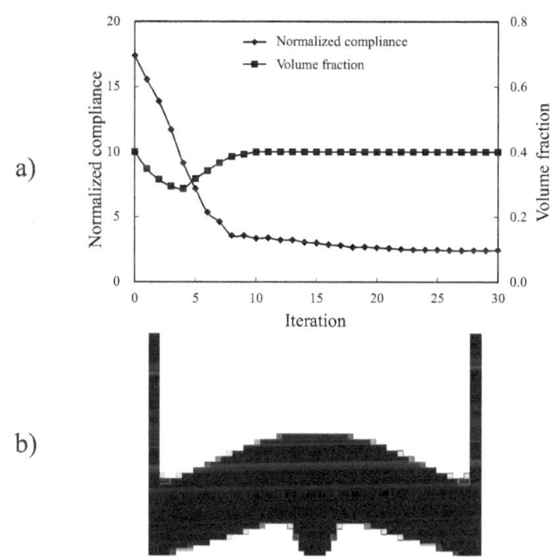


**Figure 4.6.** *Optimization design with RAMP model ($\Delta T$ = 3 K, $q_E$ = 8, $q_\beta$ = 0,): a) iteration histories of the normalized compliance and the volume fraction; b) structural configuration at last iteration*

### *Influence of the upper bound of volume constraint upon the optimization results*

With $q_E = 8$ and $q_\beta = 0$, initial values of all design variables are set to be the prescribed upper bound of volume fraction. As shown in Figure 4.7, the compliance of the optimized structure decreases monotonously and is then stabilized when $vf_U > 0.7$. Note that the normalized compliance is even less than 1 when $vf_U > 0.8$. This means that the corresponding optimized configurations are better than a fully solid structure. Furthermore, it should be highlighted that the volume constraint is active only if the upper bound is 0.965. This is just like the similar phenomenon observed in the body force problem [BRU 05].

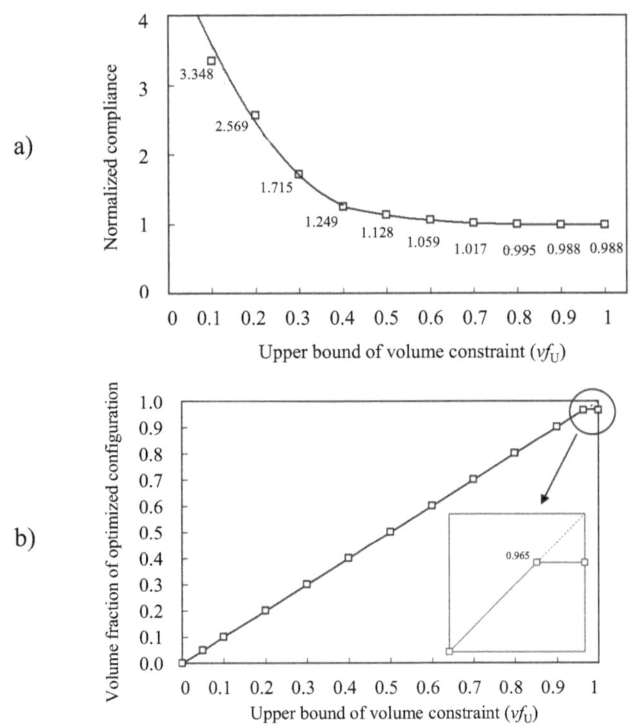


**Figure 4.7.** *Influence of the upper bound of volume constraint upon the optimization results: a) influence upon the normalized compliance; b) influence upon the volume fraction of optimized configuration*

The optimized structures are shown in Figure 4.8. The configurations have similar topologies and are in good accordance with the results of Rodrigues and Fernandes [ROD 95] for a wide range when $0.2 \leqslant vf_U \leqslant 1$. Moreover, a two-branch configuration is obtained for very small value $vf_U = 0.1$.

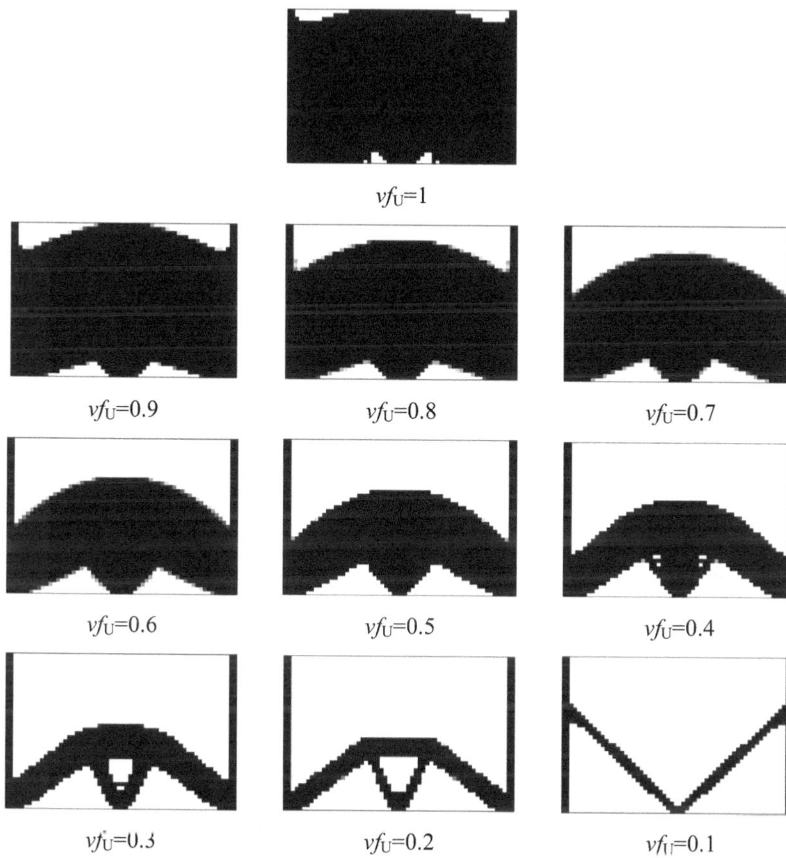


**Figure 4.8.** *Influence of the upper bound of volume constraint upon the optimal configuration*

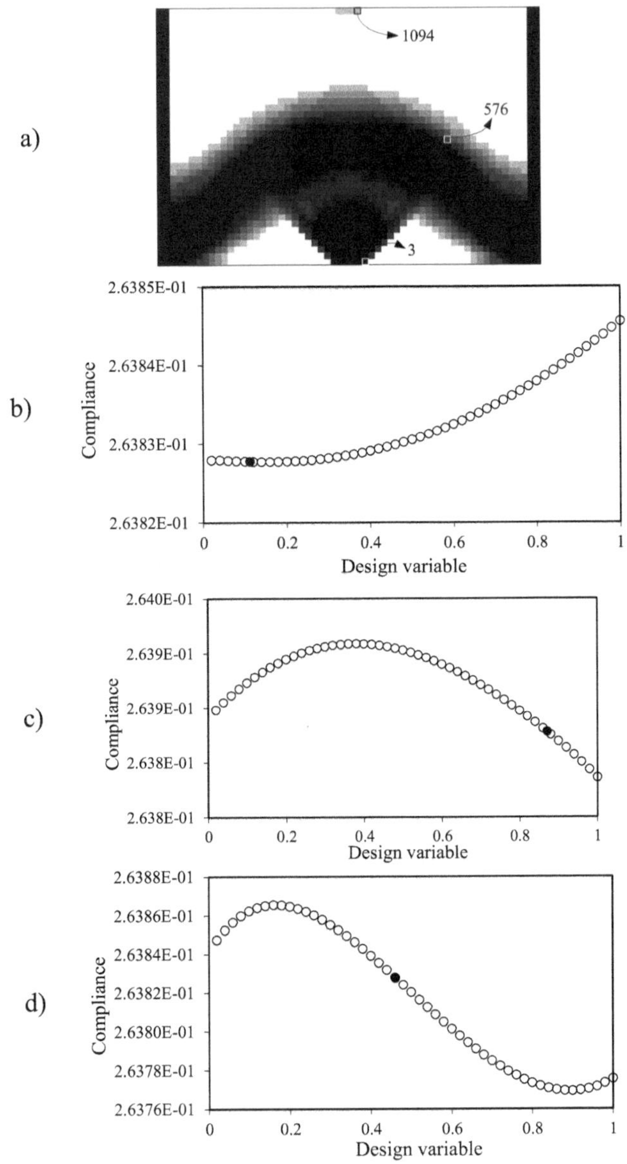


**Figure 4.9.** *Evolution curve of the compliance according to the design variable: a) optimization result after six iterations; b) evolution curve of the compliance versus the design variable (i = 1,094); c) evolution curve of the compliance versus the design variable (i = 3); d) evolution curve of the compliance versus the design variable (i = 576)*

*Discussions*

In sensitivity analysis, it is found that the sensitivity may have a non-constant sign due to the design-dependence of the thermal stress load. Actually, the thermal stress load makes the structural compliance complicated and the evolution curves of the objective function are plotted and illustrated here. Take the optimization problem of Figure 4.5 as an example and consider three elements related to the design result after six iterations shown in Figure 4.9(a). The evolution curves of the compliance versus these design variables are plotted in Figures 4.9(b–d). The solid point refers to the current solution. Obviously, the evolution curves of the compliance might be convex, concave or even non-monotonous with respect to the concerned design variable so that local solutions might exist in these optimization problems.

As is well known, concave functions will bring more difficulties into the approximation and numerical optimization. The popular optimizations, such as ConLin, MMA, GCMMA and MDQA, are all based on convex approximations and lack the ability to find out the global optimal solution. Thus, the solution presented here is not guaranteed to be the global optimal solution.

### 4.3.3.2. *Computing the rise in temperature*

As problems considered in section 4.3.3.1 are subjected to a constant temperature field, interpolation models are only applied to Young's modulus and TSC. In this section, the steady-state heat conduction is taken into account so that the interpolation model is also adopted for the thermal conductivity.

Suppose $vf_U = 0.4$ for structure 1 illustrated in Figure 4.2. Apart from the applied external force 10 kN, a fixed heat flow $Q$ is applied at the same point.

Effects of the heat flow upon the normalized optimization compliance are shown in Figure 4.10. The normalized compliance curve is approximately linear with respect to the heat flow. Corresponding optimized configurations are shown in Figure 4.11.

Noticeably, the optimized configuration with $Q = 0$ is quite different from those with $Q > 0$. It implies that the existence of the heat flow makes the material layout more complicated. Likewise, the optimized configurations are considerably different from those with a constant temperature field. Both the temperature and displacement fields of the optimal structure are illustrated in Figure 4.12.

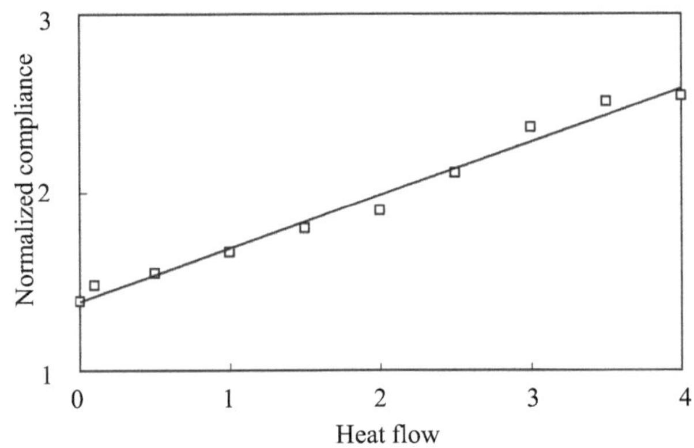


**Figure 4.10.** *Influence of the heat flow upon the normalized optimization compliance*

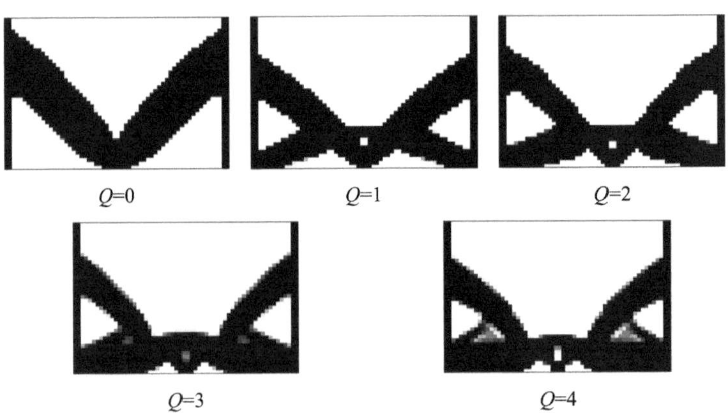


**Figure 4.11.** *Influences of the heat flow upon the optimized configuration*

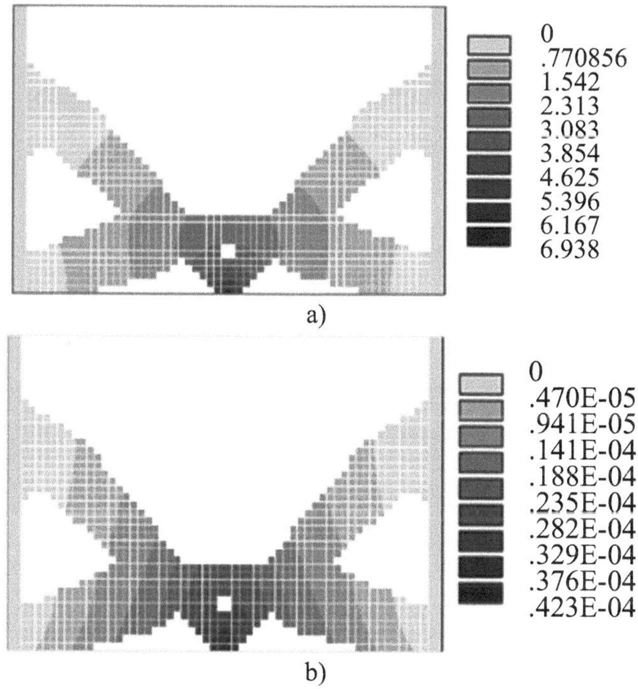


**Figure 4.12.** *Distributions of temperature and displacement fields of the optimized structure for Q = 2: a) distribution of the temperature field; b) distribution of the displacement field*

## 4.4. Thermo-elastic topology optimization with multiple materials

From the engineering viewpoint, it is a common practice to use multiphase materials for the sake of lightweight and multifunctional designs. Topology optimization with multiphase materials was first investigated by Thomsen [THO 92]. Later, typical works were focused on the extension of the SIMP/RAMP models and varieties of topology optimization problems with multiple materials including designs of microstructures with the extreme equivalent property [SIG 97], thermo-elastic problem subjected to the volume constraint [GAO 10] and multiphysics actuator design [SIG 01b]. Simultaneous design of the structural layout and discrete fiber orientation was also dealt with using an extension of the SIMP scheme, for example, the so-called discrete material optimization (DMO) scheme [STE 05], shape functions with penalization (SFP) scheme [BRU 11] and bi-value coding parameterization (BCP) scheme [GAO 12].

The ESO was also applied to address multiple materials [HUA 09]. An evolutionary approach using discrete variables was proposed to solve the mass minimization problem with multiple materials and strength constraints [RAM 11]. Alternatively, the level set method and the phase field method were applied to address topology optimization problems with multiple materials, including both the stiffness maximization problem [MEI 04a] and heat conduction problem [ZHU 10b]. It should be noted that the implicit description of the interfaces between two distinct solid material phases is the basis of this approach. Some other schemes should be mentioned here: Yin and Ananthasuresh [YIN 01] proposed a multimaterial interpolation model based on the so-called peak function; Jung and Gea [JUN 06] constructed a variable-inseparable multimaterial model for the design of an energy-absorbing structure; Yoon [YOO 11] presented the so-called patch stacking method for the nonlinear dynamic problem with multiple materials.

In the earlier work, the material amount was controlled by the volume constraint of each candidate material phase. In the engineering design sense, the volume constraint is less significant than the mass constraint to the whole structure. Although both constraints are identical when only one single solid material phase is present, the situation changes completely in the case of multiple materials due to the differences in material densities. The mass constraint of multiple materials was investigated under pure mechanical loads for the structural compliance minimization [GAO 11]. Two interpolation schemes, namely, recursive multiphase materials interpolation (RMMI) and uniform multiphase materials interpolation (UMMI) were discussed. It was demonstrated that the mass constraint is more beneficial than the volume constraint in the sense that the structural stiffness can be further increased and multiple material properties can be automatically matched.

This section focuses on topology optimization with multiple materials subject to the mass constraint and both mechanical and thermal loading are taken into account.

### 4.4.1. *Standard optimization formulation*

In this section, topology optimization with multiple materials is stated as

$$\begin{array}{ll} \text{find} & \boldsymbol{\eta}=\{\eta_{ij}\} \quad i=1,2,\ldots,n_{\mathrm{e}};\ j=1,2,\ldots,m \\ \min & C=\mathbf{F}^{\mathrm{T}}\mathbf{U} \\ \text{s.t} & \mathbf{F}=\mathbf{KU} \\ & \mathbf{Q}=\mathbf{HT} \\ & M\le M_{\mathrm{U}} \\ & 0<\eta_{\mathrm{L}}\le\eta_{ij}\le 1 \end{array} \qquad [4.18]$$

in which $m$ is the number of candidate material phases and $\eta_{ij}$ represents the presence (1) or absence (0) of the $j$th candidate material in the $i$th finite element. The structural mass $M$ should be less than its upper bound $M_{\mathrm{U}}$.

### 4.4.2. *Sensitivity analysis*

In a design problem involving both mechanical and thermal loads, the elasticity matrix and TSC should be parameterized to compute the element stiffness matrix and thermal stress load vectors, respectively. If a heat conduction analysis is required to compute the temperature field, the heat conductivity coefficient $\kappa_i$ should be parameterized as well. Generally, these parameterized material properties can be expressed as the weighted summation of all of the candidate material phases.

$$\mathbf{D}_i=\sum_{j=1}^{m} w_{ij}^{D}\mathbf{D}^{(j)} \quad \beta_i=\sum_{j=1}^{m} w_{ij}^{\beta}\beta^{(j)} \quad \kappa_i=\sum_{j=1}^{m} w_{ij}^{\kappa}\kappa^{(j)} \qquad [4.19]$$

Herein, the superscript ($j$) denotes the $j$th candidate material phase. The TSC of the $j$th material phase corresponds to $\beta^{(j)} = E^{(j)}\alpha^{(j)}/(1\text{-}2\mu^{(j)})$.

In [GAO 10], the UMMI was found to be superior to the RMMI because the former made it possible to formulate the mass constraint in a linear form with separable design variables. One such formulation benefits the problem resolution by means of mathematical programming approaches, specifically convex programming methods. Hence, the RAMP scheme is utilized in combination with the UMMI scheme known as DMO in the work of Stegmann and Lund [STE 05]. The weighting functions in the above parameterization models then correspond to

$$
\begin{aligned}
w_{ij}^{D} &= R_{ij}^{D}\prod_{\substack{\xi=1\\ \xi\neq j}}^{m}\left(1-R_{i\xi}^{D}\right) & R_{ij}^{D} &= \frac{\eta_{ij}}{1+q^{D}\left(1-\eta_{ij}\right)}\\
w_{ij}^{\beta} &= R_{ij}^{\beta}\prod_{\substack{\xi=1\\ \xi\neq j}}^{m}\left(1-R_{i\xi}^{\beta}\right) & R_{ij}^{\beta} &= \frac{\eta_{ij}}{1+q^{\beta}\left(1-\eta_{ij}\right)}\\
w_{ij}^{\kappa} &= R_{ij}^{\kappa}\prod_{\substack{\xi=1\\ \xi\neq j}}^{m}\left(1-R_{i\xi}^{\kappa}\right) & R_{ij}^{\kappa} &= \frac{\eta_{ij}}{1+q^{\kappa}\left(1-\eta_{ij}\right)}
\end{aligned}
\qquad [4.20]
$$

where $R_{ij}^{D}$, $R_{ij}^{\beta}$ and $R_{ij}^{\kappa}$ are calculated using $q^{D}$, $q^{\beta}$ and $q^{\kappa}$ which are internal parameters in the RAMP scheme.

By combining the UMMI and RAMP schemes, the sensitivity of the structural compliance then corresponds to

$$
\frac{\partial C}{\partial \eta_{ij}} = 2\mathbf{U}^{\mathrm{T}}\left(\frac{\partial \mathbf{F}^{\mathrm{m}}}{\partial \eta_{ij}}+\frac{\partial \mathbf{F}^{\mathrm{th}}}{\partial \eta_{ij}}\right)-\mathbf{U}^{\mathrm{T}}\frac{\partial \mathbf{K}}{\partial \eta_{ij}}\mathbf{U} = 2\mathbf{U}_{i}^{\mathrm{T}}\frac{\partial \mathbf{F}_{i}^{\mathrm{th}}}{\partial \eta_{ij}}-\mathbf{U}_{i}^{\mathrm{T}}\frac{\partial \mathbf{K}_{i}}{\partial \eta_{ij}}\mathbf{U}_{i} \qquad [4.21]
$$

Evidently, $\partial\mathbf{K}i/\partial\eta_{ij}$ and $\partial\mathbf{F}_{i}^{\mathrm{th}}/\partial\eta_{ij}$ can be easily derived at the element level. Mathematically, $\partial\mathbf{K}i/\partial\eta_{ij}$ is written as

$$\frac{\partial \mathbf{K}_i}{\partial \eta_{ij}} = \int_{V_i} \mathbf{B}_i^{\mathrm{T}} \frac{\partial \mathbf{D}_i}{\partial \eta_{ij}} \mathbf{B}_i \, dV = \sum_{\varsigma=1}^{m} \left( \frac{\partial w_{i\varsigma}^D}{\partial \eta_{ij}} \mathbf{K}_i^{(\varsigma)} \right)$$

$$\mathbf{K}_i^{(\varsigma)} = \int_{V_i} \mathbf{B}_i^{\mathrm{T}} \mathbf{D}^{(\varsigma)} \mathbf{B}_i \, dV$$

$$\frac{\partial w_{i\varsigma}^D}{\partial \eta_{ij}} = \begin{cases} \dfrac{\partial R_{ij}^D}{\partial \eta_{ij}} \prod\limits_{\substack{\xi=1 \\ \xi \neq j}}^{m} \left(1 - R_{i\xi}^D\right) & j = \varsigma \\ -R_{i\varsigma}^D \dfrac{\partial R_{ij}^D}{\partial \eta_{ij}} \prod\limits_{\substack{\xi=1 \\ \xi \neq j, \xi \neq \varsigma}}^{m} \left(1 - R_{i\xi}^D\right) & j \neq \varsigma \end{cases} \qquad [4.22]$$

$$\frac{\partial R_{ij}^D}{\partial \eta_{ij}} = \frac{1 + q^D}{\left(1 + q^D \left(1 - \eta_{ij}\right)\right)^2}$$

A constant temperature rise ($\partial \Delta T_i / \partial \eta_{ij} = 0$) is assumed and the partial derivative of the thermal stress load is stated as

$$\frac{\partial \mathbf{F}_i^{\mathrm{th}}}{\partial \eta_{ij}} = \sum_{\varsigma=1}^{m} \left( \frac{\partial w_{i\varsigma}^\beta}{\partial \eta_{ij}} \beta^{(\varsigma)} \right) \overline{\mathbf{F}}_i^{\mathrm{th}}$$

$$\frac{\partial w_{i\varsigma}^\beta}{\partial \eta_{ij}} = \begin{cases} \dfrac{\partial R_{ij}^\beta}{\partial \eta_{ij}} \prod\limits_{\substack{\xi=1 \\ \xi \neq j}}^{m} \left(1 - R_{i\xi}^\beta\right) & j = \varsigma \\ -R_{i\varsigma}^\beta \dfrac{\partial R_{ij}^\beta}{\partial \eta_{ij}} \prod\limits_{\substack{\xi=1 \\ \xi \neq j, \xi \neq \varsigma}}^{m} \left(1 - R_{i\xi}^\beta\right) & j \neq \varsigma \end{cases} \qquad [4.23]$$

$$\frac{\partial R_{ij}^\beta}{\partial \eta_{ij}} = \frac{1 + q^\beta}{\left(1 + q^\beta \left(1 - \eta_{ij}\right)\right)^2}$$

According to the sensitivity analysis, $\partial C / \partial \eta_{ij}$ is positive or negative depending on the relative magnitudes of the partial derivatives of the stiffness matrix and thermal stress load. This implies that the compliance formulation might be non-monotonic.

### 4.4.3. *Mass constraint*

#### 4.4.3.1. *Mass constraint formulation*

The formulation of the mass constraint depends on how the densities of the candidate material phases are interpolated. First, consider the linear formulation [GAO 11] with the density of element $i$ interpolated as

$$\rho_i = \sum_{j=1}^{m} \eta_{ij} \rho^{(j)} \tag{4.24}$$

The corresponding mass constraint of multiple materials then reads

$$M_{\mathrm{L}} = \sum_{i=1}^{n_{\mathrm{e}}} \sum_{j=1}^{m} \eta_{ij} \rho^{(j)} V_i \leq M_{\mathrm{U}} \tag{4.25}$$

Obviously, the linear form and separability of the design variables are favorable to the optimization procedure. The sensitivity of the mass constraint can easily be derived as

$$\frac{\partial M_{\mathrm{L}}}{\partial \eta_{ij}} = \rho^{(j)} V_i \tag{4.26}$$

In this section, an alternative formulation of the density using the UMMI scheme is also studied,

$$\rho_i = \sum_{j=1}^{m} \left( \eta_{ij} \prod_{\substack{\xi=1 \\ \xi \neq j}}^{m} \left(1 - \eta_{i\xi}\right) \right) \rho^{(j)} \tag{4.27}$$

The corresponding mass constraint is then written as

$$M_{\mathrm{N}} = \sum_{i=1}^{n_{\mathrm{e}}} \sum_{j=1}^{m} \left( \eta_{ij} \prod_{\substack{\xi=1 \\ \xi \neq j}}^{m} \left(1 - \eta_{i\xi}\right) \right) \rho^{(j)} V_i \leq M_{\mathrm{U}} \tag{4.28}$$

and the sensitivity is expressed as

$$\frac{\partial M_{\mathrm{N}}}{\partial \eta_{ij}} = \left( \prod_{\substack{\xi=1 \\ \xi \neq j}}^{m} (1-\eta_{i\xi}) \rho^{(j)} - \sum_{\substack{\varsigma=1 \\ \varsigma \neq j}}^{m} \left( \eta_{i\varsigma} \prod_{\substack{\xi=1 \\ \xi \neq j, \xi \neq \varsigma}}^{m} (1-\eta_{i\xi}) \right) \rho^{(\varsigma)} \right) V_i \quad [4.29]$$

In numerical tests, the nonlinear mass constraint is found to be frequently violated, which will terminate the optimization process. Therefore, $M_{\mathrm{L}}$ is used instead of $M_{\mathrm{N}}$ in this work.

In comparison, the volume constraint related to multiple materials is also presented below. For each candidate material phase $j$, the volume constraint is written as

$$V_{\mathrm{C}} = \sum_{i=1}^{n_e} V_i \eta_{ij} \leq V_{\mathrm{U}j} = vf_{\mathrm{U}j} \cdot V \quad j = 1, \ldots, m \quad [4.30]$$

Herein, $V_{\mathrm{U}j}$ denotes the prescribed upper bound of the volume constraint of the $j$th candidate material. The upper bound of the volume fraction $vf_{\mathrm{U}j}$ is defined as the division of $V_{\mathrm{U}j}$ by the summation of all designable elements $V$, namely, $vf_{\mathrm{U}j} = V_{\mathrm{U}j}/V$. Obviously, $0<vf_{\mathrm{U}j}<1$ for each meaningful volume constraint. Usually, the number of volume constraints equals the number of candidate material phases $m$ and $\sum_{j=1}^{m} vf_{\mathrm{U}j} \leq 1$ for all volume constraints. The sensitivity of the volume constraint is obviously constant,

$$\frac{\partial V_{\mathrm{C}}}{\partial \eta_{ij}} = V_i \quad [4.31]$$

### 4.4.3.2. *Relationship between volume and mass constraints*

Without a loss of generality, suppose that an arbitrary feasible solution $\eta = \{\eta_{ij}\}$ exists in the sense that the set of volume constraints is satisfied. Based on the condition $0 \leq 1-\eta_{ij}<1$, it follows that

$$\sum_{i=1}^{n_e}\left(\eta_{ij}\prod_{\substack{\xi=1\\ \xi\neq j}}^{m}\left(1-\eta_{i\xi}\right)\right)V_i \leq \sum_{i=1}^{n_e}\eta_{ij}V_i \leq V_{Uj} \qquad [4.32]$$

The multiplication of this term by the density of the material phase $j$, $\rho^{(j)}$, gives rise to

$$\sum_{i=1}^{n_e}\left(\eta_{ij}\prod_{\substack{\xi=1\\ \xi\neq j}}^{m}\left(1-\eta_{i\xi}\right)\right)V_i\rho^{(j)} \leq \sum_{i=1}^{n_e}\eta_{ij}V_i\rho^{(j)} \leq V_{Uj}\rho^{(j)} \qquad [4.33]$$

Then, the summation of all solid materials results in

$$\sum_{j=1}^{m}\sum_{i=1}^{n_e}\left(\eta_{ij}\prod_{\substack{\xi=1\\ \xi\neq j}}^{m}\left(1-\eta_{i\xi}\right)\right)V_i\rho^{(j)} \leq \sum_{j=1}^{m}\sum_{i=1}^{n_e}\eta_{ij}V_i\rho^{(j)} \leq \sum_{j=1}^{m}V_{Uj}\rho^{(j)} \qquad [4.34]$$

Now, suppose that the structure mass is specified as the upper bound $M_U$, we then have

$$\sum_{j=1}^{m}V_{Uj}\rho^{(j)} = M_U \qquad [4.35]$$

The combination of equation [4.34] and equation [4.35] yields the linear and nonlinear forms of the mass constraint given in equation [4.25] and equation [4.28], respectively. This means that an arbitrary feasible solution of the volume-constrained optimization problem is always within the feasible domain of the mass-constrained design problem.

It concludes that the following relation theoretically exists for design spaces defined by the volume and mass constraints.

$$\Omega_V \subseteq \Omega_{M_L} \subseteq \Omega_{M_N} \qquad [4.36]$$

Geometrically, this means that under the precondition of the same mass of the structure, the volume constraint defines a subdesign space of the linear mass constraint, while the latter defines a subdesign space of the nonlinear mass constraint. Therefore, the mass constraint always produces a stiffer configuration than the volume constraint in the compliance minimization sense.

At the same time, the following relation holds:

$$M_{\mathrm{N}} \leq M_{\mathrm{L}} \qquad [4.37]$$

Note that $M_{\mathrm{N}} = M_{\mathrm{L}}$ only if $\eta_{ij} = 1$ and $\eta_{i\xi} = 0$ ($\xi \neq j$) with one single material or $\eta_{ij} = 0$ with a void in each element. Although $M_{\mathrm{N}}$ is not adopted in the optimization formulation, values of $M_{\mathrm{N}}$ and $M_{\mathrm{L}}$ can be utilized to check the status of the presence or absence of mixed material.

## 4.4.4. *Improved optimization formulation*

In this section, a typical numerical example is tested to illustrate the mixed material status if the presented standard optimization formulation is used. Then, an improved optimization formulation is proposed to overcome the resulting defect.

### 4.4.4.1. *Mixed material status*

Consider the biclamped plane structure with two non-designable domains (the dark areas) on both sides, as shown in Figure 4.2. The upper bound of the mass constraint is $M_{\mathrm{U}} = 10$. A vertical force $F^{\mathrm{m}}$ = 8kN is applied, and the reference temperature for the thermal strain calculation is 20°C. Suppose two isotropic solid material phases are available (black for M2 and gray for M1) with their properties listed in Table 4.1. Note that M2 has large values of the Young's modulus, stiffness-to-density ratio and TSC. This implies that M2 is superior to M1 in bearing mechanical loads, but the produced thermal stress load from M2 does not favor the reduction of structural compliance. According to [GAO 11, STE 05], a uniform initial weighting in the UMMI scheme is helpful for topology optimization problems under pure mechanical loads to avoid the local optimum close to the initial point. Therefore, uniform initial design variables ($\eta_{i1} = \eta_{i2} = 0.05$) are adopted here as a feasible starting point.

| Material phase | M1 (TC4) | M2 (18Mn2CrMoBA) |
|---|---|---|
| Density (kg/m$^3$) | 4,440 | 7,850 |
| Young's modulus (GPa) | 105 | 190 |
| Thermal conductivity (W/(m ·°C)) | 8.7 | 35.38 |
| Poisson's ratio | 0.34 | 0.28 |
| Coefficient of thermal expansion (×10$^{-6}$/°C) | 9.10 | 12.4 |
| Thermal stress coefficient (MPa/°C) | 2.986 | 5.355 |

**Table 4.1.** *Basic properties of two solid material phases*

Suppose a uniform temperature of 100 K is applied over the whole structure, the optimization results are illustrated in Figure 4.13. It is found that $M_N<M_L$, even after 80 iterations, which means that the mixed material status still exists in some elements. Consider elements 3 and 33 as examples, with the evolution histories of their design variables plotted in Figures 4.13(c) and (d). $\eta_{i2} = 0$ after tens of iterations, while $\eta_{i1}$ approaches 0.039 and 0.177, respectively, for elements 3 and 33. Of particular interest, the evolution track is self-intersecting for element 33.

#### 4.4.4.2. *Optimization formulation with an artificial penalty term*

To avoid the mixed material status of each element, we introduce the following artificial penalty term

$$\delta = \sum_{i=1}^{n_e} \left( \sum_{\xi=1}^{m} \sum_{\zeta=\xi+1}^{m} \eta_{i\xi} \eta_{i\zeta} \right) \qquad [4.38]$$

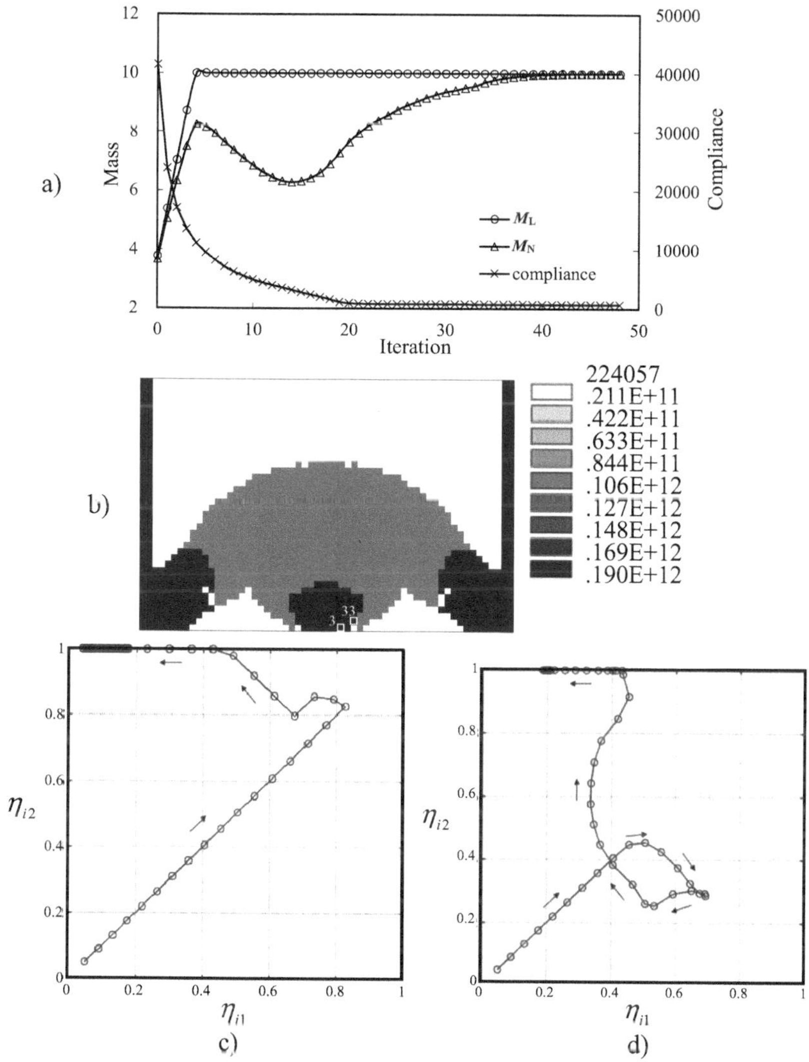


**Figure 4.13.** *Optimization results under uniform $\Delta T$ = 80 K: a) iteration histories of the compliance and mass; b) distribution of Young's modulus; c) evolution track (i = 3); d) evolution track (i = 33)*

The optimization formulation is then rewritten as

$$\begin{aligned} \text{find} \quad & \boldsymbol{\eta} = \{\eta_{ij}\} \quad i = 1,2,\ldots,n_e;\ j = 1,\ldots,m \\ \text{min} \quad & C + a \cdot \delta \\ \text{s.t} \quad & \mathbf{F}^{\text{th}} + \mathbf{F}^{\text{m}} = \mathbf{KU} \\ & \mathbf{Q} = \mathbf{HT} \\ & M \le M_{\text{U}} \\ & 0 < \eta_{\text{L}} \le \eta_{ij} \le 1 \end{aligned} \qquad [4.39]$$

Here, $\delta$ acts as a positive penalty to force the design variables in each element toward a solid ($\eta_{ij} = 1$ and $\eta_{i\xi} = 0$ ($\xi \neq j$)) or void ($\eta_{ij} = 0$) material status. This situation occurs at $\delta = 0$ so that the objective function in equation [4.39] equals the structural compliance. Note that $a = \tilde{a}C_0/\delta_0$ is a scale parameter defined to avoid large differences between $\delta$ and $C$. $\delta_0$ and $C_0$ are the initial values.

With the updated formulation in equation [4.39], the biclamped rectangular structure is optimized. The influence of parameter $\tilde{a}$ upon the optimization process is first studied. As shown in Figure 4.14(a), the iterations of $\delta$ oscillate for $\tilde{a} = 0$ and $\tilde{a} = 0.04$, while large values of $\tilde{a}$ lead to stable iterations. In Figure 4.14(b), it is shown that a large value of $\tilde{a}$ will push $\delta$ toward zero, but increases the structural compliance. In summary, a small value for the parameter $\tilde{a}$ leads to stiff configurations with mixed material status, while large values of $\tilde{a}$ yield clear but weak configurations.

A treatment using variable $\tilde{a}$ is proposed to seek clear and stiff configurations where $\tilde{a}$ is gradually increased from 0.01 to 0.1 in 40 iterations. The $\delta$ stabilizes as the iteration count increases, as plotted in Figure 4.14(a). The optimization results are shown in Figure 4.15. Clearly, the iteration processes of the compliance and material costs are globally stable. Linear and nonlinear formulations of the mass constraint are used to evaluate the effect of the artificial penalty term. Because the masses of the linear and nonlinear forms converge and are nearly identical after 40 iterations, the value of $\delta$ tends to zero. Consequently, a clear configuration is obtained without mixed material status. The optimization formulation in equation [4.39] is thus adopted in the following numerical tests instead of the original one that was used in equation [4.18].

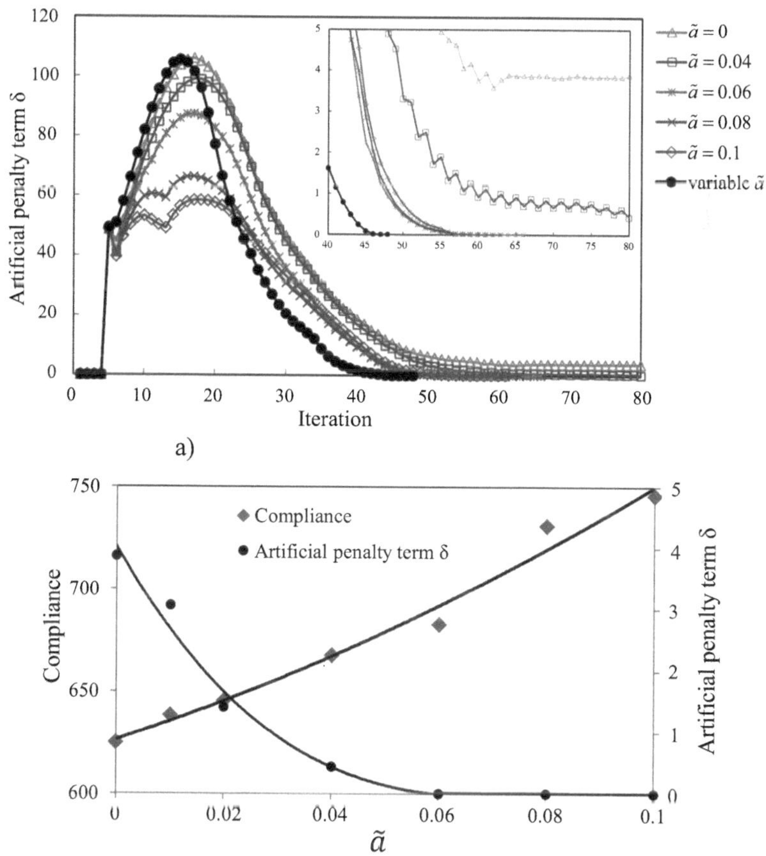


**Figure 4.14.** *Influence of ã upon the optimization process: a) influence of ã on the iteration histories of δ; b) influence of ã on the compliance and δ of the optimized results. For a color version of this figure, see www.iste.co.uk/zhang/topology.zip*

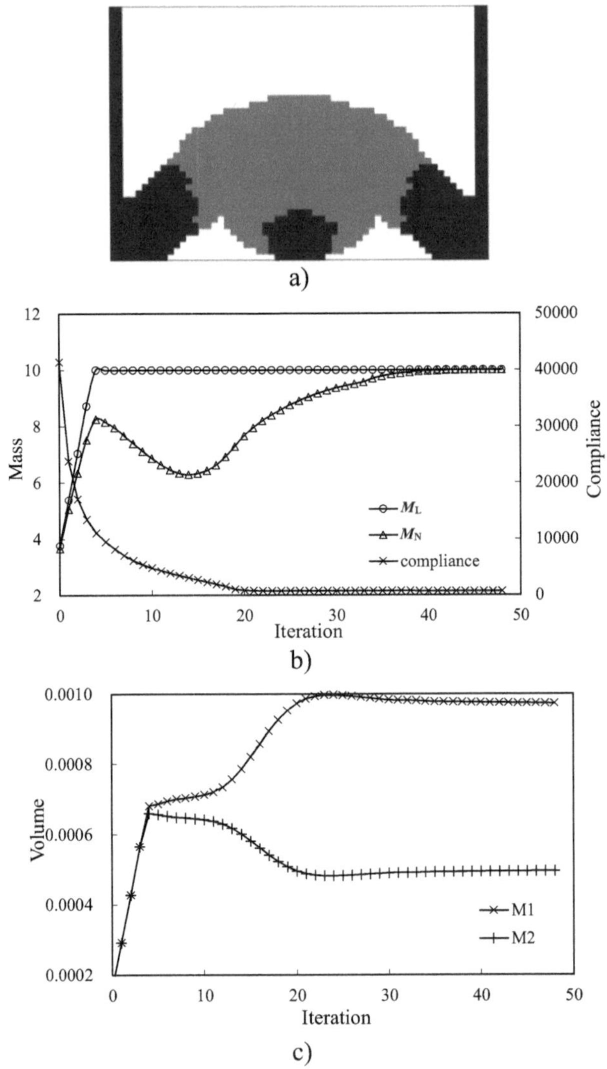


**Figure 4.15.** *Optimization results with equation [4.39] under uniform $\Delta T$ = 80 K: a) optimized configuration (C = 662.4 $\delta = 0.44\times10^{-2}$); b) iteration histories of the compliance and mass; c) iteration histories of the volumes of two solid materials*

To gain further understanding of the nonlinearity and complexity of the investigated problem, the intermediate design result after

17 iterations is shown in Figure 4.16(a). The distribution of the compliance value versus the design variables of element 61 is illustrated in Figure 4.16(b). Obviously, the compliance is non-monotonic with respect to the concerned design variables. Particularly, there exists a minor peak around (0.4, 0) and a cliff near (1, 1).

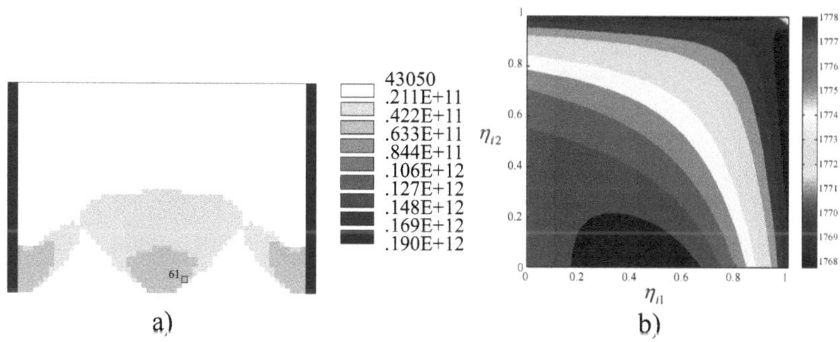


**Figure 4.16.** *Distribution of the compliance value versus design variables of element 61: a) Distribution of Young's modulus after 17 iterations; b) Compliance versus design variables (i = 61). For a color version of this figure, see www.iste.co.uk/zhang/topology.zip*

### 4.4.5. *Numerical examples*

More numerical examples are tested using the improved optimization formulation in equation [4.39] with the artificial penalty term and variable $\tilde{a}$, which is gradually increased from 0.01 to 0.1 in 40 iterations.

#### *2D structure*

The biclamped plane structure is further tested. The influence of the temperature rise on the optimization results is also investigated. Suppose the mechanical force remains unchanged and the structural mass is constrained by $M_U = 10$ in all tests. Figure 4.17(a) shows that the compliance increases with the rise of uniform temperature due to the increase in thermal stress loads. Figure 4.17(b) shows that the amount of M1 increases with the temperature rise because of its small TSC value. Some voids thus occur inside the structure to allow for thermal stress release. In the extreme case of either a small or absent temperature rise, only M2 is used due to its large Young's modulus value. Therefore, the optimized configuration result is a compromise between mechanical and thermal stress loads.

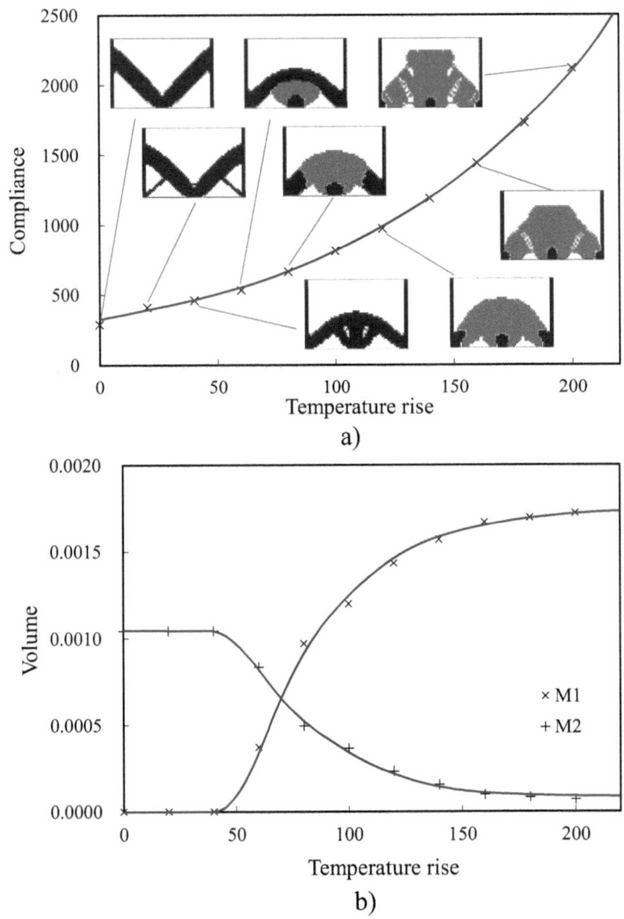


**Figure 4.17.** *Influence of the temperature rise on the optimized result at $M_U = 10$: a) influence of the temperature rise on the compliance; b) influence of the temperature rise on the volume of each candidate material phase*

Now, the effect of the upper bound of the mass constraint on the optimized results is investigated for a uniform temperature rise (80 K).

Figure 4.18 indicates that the structural compliance of the optimized configuration naturally decreases with the increase in $M_U$. The mass constraint reaches its upper bound in all tests. If the mass is strongly limited by a small value of $M_U$, the solution corresponds to a two-branch structure consisting of only M2. In contrast, a large value of $M_U$ results in similar optimized configurations that are filled with both materials. In detail, M2 is placed around the mechanical loads and lower corners of the non-designable area. M1 transfers loads to both vertical edges. The compliances are almost the same for $M_U \geqslant 8$ despite differences in both the amount of each available material and the optimized configurations.

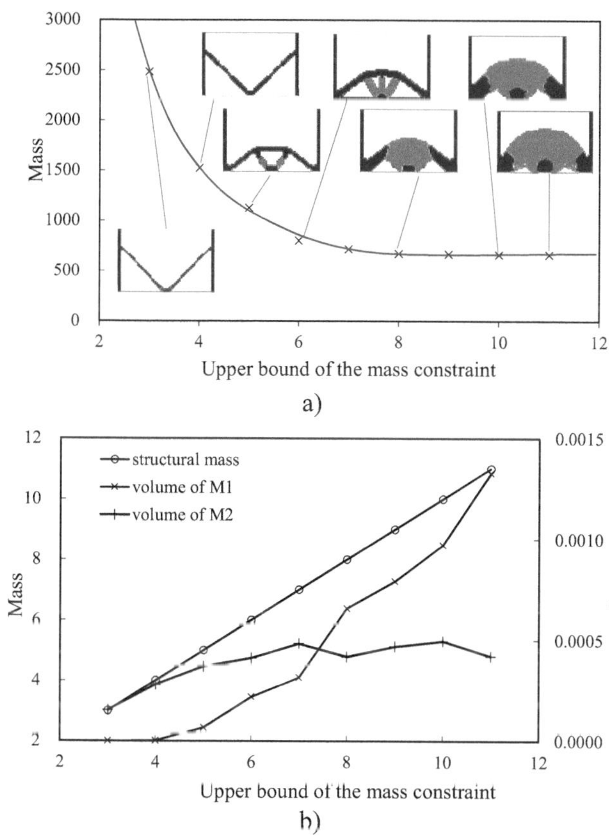


**Figure 4.18.** *Influence of the upper bound of the mass constraint on the optimized result*

Furthermore, the volume and mass constraints are compared. To do this, two volume constraints are introduced for M1 and M2 to replace the linear mass constraint. The upper bound of each volume constraint is correspondingly defined to match $M_U = 10$. As shown in Figure 4.19, the optimized configuration with the mass constraint is found to be the stiffest whatever the upper bounds of the volume constraints are. This situation theoretically confirms the benefit of using the mass constraint over the volume constraint according to equation [4.36].

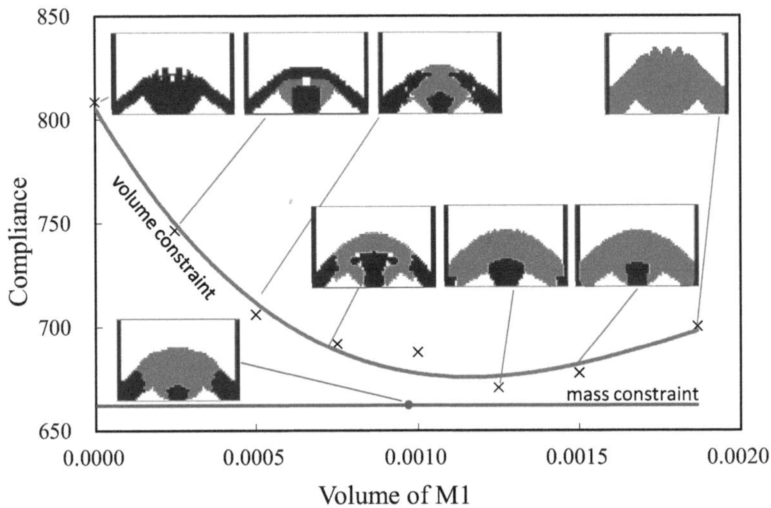


**Figure 4.19.** *Comparisons of optimized results under mass or volume constraints with the same $M_U = 10$*

Now, the effects of the initial values of the design variables are highlighted in the case of multiple materials. Tests are conducted with fixed $\eta_{i2} = 0.05$ and different $\eta_{i1}$ as initial values. Figure 4.20(a) indicates that the resulting curve is non-monotonic. Uniform initial weights are not the absolute best choice, although they are suggested to address problems under a design-independent mechanical load. Figure 4.20(b) indicates that the amount of M2 obviously decreases with the increase in the initial value of $\eta_{i1}$. In these tests, the mass constraint always remains active.

According to [GAO 11] and [STE 05], a uniform initial weighting in the UMMI scheme is desirable for topology optimization problems under pure mechanical loads to avoid the local optimum. In this work, due to the presence of a design-dependent thermal stress load, the non-uniform initial weighting might yield a local optimum, which is better than the uniform initial weighting.

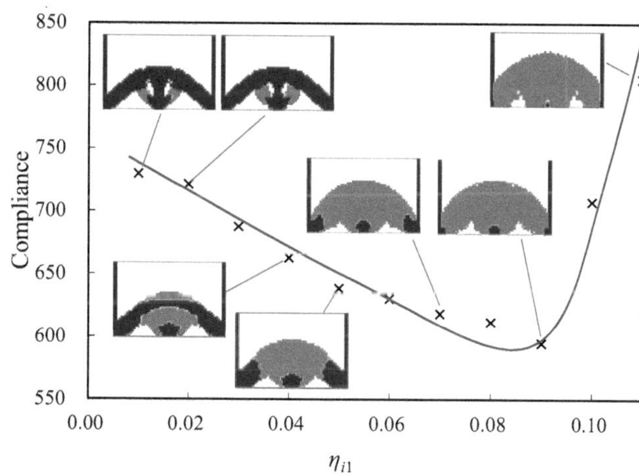


a)

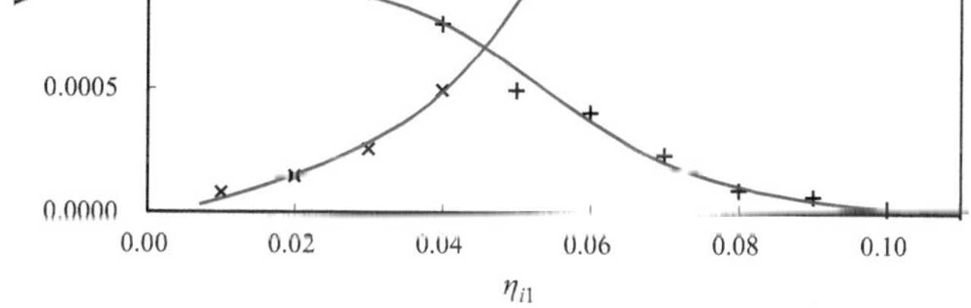


b)

**Figure 4.20.** *Influence of the initial value of $\eta_{i1}$ on the optimized result ($\eta_{i2}$ = 0.05); a) influence of the initial value of $\eta_{i1}$ on the compliance; b) influence of the initial value of $\eta_{i1}$ on the volume of each candidate material phase*

### *3D structure*

The proposed method is now tested for a large-scale engineering problem. The structure used is shown in Figure 4.21. It is fixed on the ends of two horizontal arms. Herein, two candidate materials, M1 and M2, are available. The structure undergoes a temperature rise and mechanical load simultaneously. The latter consists of pressure applied to the inner surface of the hole with a cosine distribution function over $\theta$ = 120° (the peak value $p_1$ = 50 MPa) and uniform pushing force ($p_2$ = 8 MPa). Suppose the loading area is a non-designable one filled with material phase M2. The reference temperature for the thermal strain calculations is assumed to be 20°C. The whole structure is meshed into 39,352 solid elements and the design variables are uniformly set to initial values of $\eta_{i1} = \eta_{i2} = 0.05$, which is a feasible starting point.

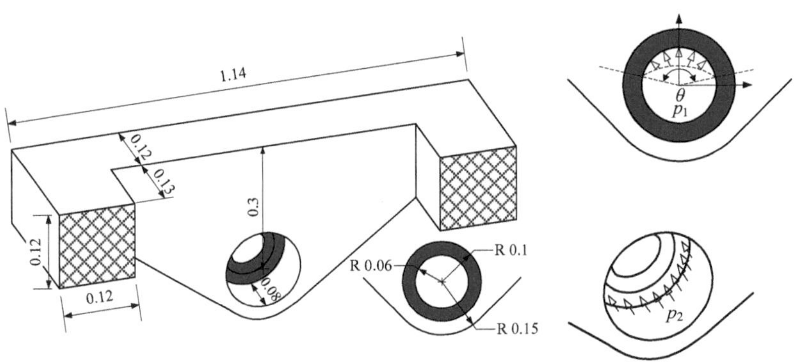


**Figure 4.21.** *Structure 2 (unit: m)*

At $M_U$ = 40, the optimized results are illustrated in Figure 4.22 (red for M2 and cyan for M1). In the case of pure mechanical loads, the whole structure consists of only M2 due to its high Young's modulus value, as shown in Figure 4.22(a). If the uniform temperature rise ($\Delta T$ = 60 K) is applied, the optimized structure shown in Figure 4.22(b) is obviously different. The structure is mainly composed of M1 due to its small TSC value, and M2 is distributed around clamped areas. A non-uniform temperature rise is tested as well. Temperatures on the inner surface of the hole and the upper surface of the structure are set to 110 and 50°C, respectively. In each iteration, a steady-state heat conduction analysis is carried out to obtain the temperature field over the whole structure. The optimized configuration and its temperature field are illustrated in Figure 4.22(c). Note that the mass constraint reaches its upper bound in all test cases.

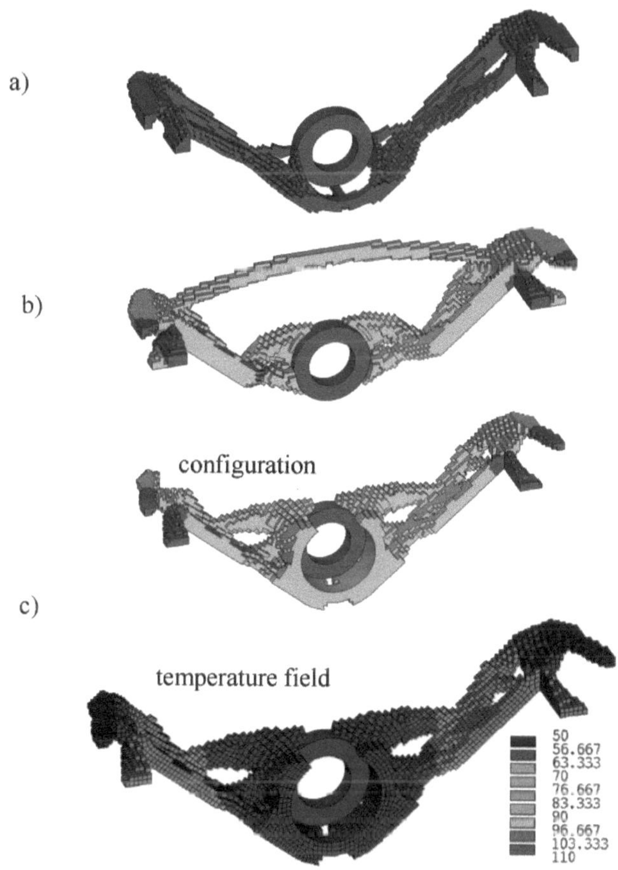


**Figure 4.22.** *Optimized results for the 3D structure: a) $\Delta T$ = 0°C (M2: 5.095 × $10^{-3}$ $m^3$); b) uniform temperature rise ($\Delta T$ = 60°C) (M1: 6.051 × $10^{-3}$ $m^3$, M2:1.673 × $10^{-3}$ $m^3$); c) non-uniform temperature rise (M1: 6.199 × $10^{-3}$ $m^3$, M2: 1.589 × $10^{-3}$ $m^3$). For a color version of this figure, see www.iste.co.uk/zhang/topology.zip*

## 4.5. Distinction between mean compliance and elastic strain energy

In all the above work, the mean compliance is used as the objective function. When only mechanical loads exist, the mean compliance and strain energy are identical. However, as the thermal loading is design-dependent, it contributes to each objective formulation differently so that both metrics are no longer the same. In fact, the thermal load follows the rule of "no material, no thermal load" in the optimization process, while the mechanical load is a fixed one. To study each formulation, the effects of mechanical and thermal loads upon the structural compliance and strain energy are investigated in detail. As shown later, the difference between both metrics results in a big difference between the optimized configurations as the thermal load increases. In this section, single material is the candidate.

### 4.5.1. *Formulations of mean compliance and elastic strain energy*

The discrete form of the mean compliance $C$ of an elastic body is written as:

$$\begin{aligned} C &= \tfrac{1}{2}\mathbf{U}^{\mathrm{T}}\mathbf{K}\mathbf{U} = \tfrac{1}{2}(\mathbf{F}^{\mathrm{m}} + \mathbf{F}^{\mathrm{th}})^{\mathrm{T}}\mathbf{U} = \tfrac{1}{2}\mathbf{U}^{\mathrm{T}}\mathbf{F}^{\mathrm{th}} + \tfrac{1}{2}\mathbf{U}^{\mathrm{T}}\mathbf{F}^{\mathrm{m}} \\ &= \tfrac{1}{2}(\mathbf{F}^{\mathrm{m}})^{\mathrm{T}}\mathbf{K}^{-1}\mathbf{F}^{\mathrm{m}} + \tfrac{1}{2}(\mathbf{F}^{\mathrm{th}})^{\mathrm{T}}\mathbf{K}^{-1}\mathbf{F}^{\mathrm{th}} + (\mathbf{F}^{\mathrm{m}})^{\mathrm{T}}\mathbf{K}^{-1}\mathbf{F}^{\mathrm{th}} \end{aligned} \qquad [4.40]$$

It should be noted that a coefficient 1/2 is added in the mean compliance formulation in order to conveniently compare with the elastic strain energy defined below. The mean compliance can be regarded as a combination of three terms: mechanical, thermal and coupled.

$$\begin{aligned} C &= C^{\mathrm{m}} + C^{\mathrm{th}} + C^{\mathrm{mth}} \\ C^{\mathrm{m}} &= \tfrac{1}{2}(\mathbf{F}^{\mathrm{m}})^{\mathrm{T}}\mathbf{K}^{-1}\mathbf{F}^{\mathrm{m}} = \tfrac{1}{2}(\mathbf{F}^{\mathrm{m}})^{\mathrm{T}}\mathbf{U}^{\mathrm{m}} \\ C^{\mathrm{th}} &= \tfrac{1}{2}(\mathbf{F}^{\mathrm{th}})^{\mathrm{T}}\mathbf{K}^{-1}\mathbf{F}^{\mathrm{th}} = \tfrac{1}{2}(\mathbf{F}^{\mathrm{th}})^{\mathrm{T}}\mathbf{U}^{\mathrm{th}} \\ C^{\mathrm{mth}} &= (\mathbf{F}^{\mathrm{m}})^{\mathrm{T}}\mathbf{K}^{-1}\mathbf{F}^{\mathrm{th}} = (\mathbf{F}^{\mathrm{m}})^{\mathrm{T}}\mathbf{U}^{\mathrm{th}} = (\mathbf{F}^{\mathrm{th}})^{\mathrm{T}}\mathbf{U}^{\mathrm{m}} \end{aligned} \qquad [4.41]$$

Physically, $C^{\mathrm{mth}}$ refers to the coupled effect between thermal and mechanic loads. It can be positive or negative depending upon the relative direction between the displacement and load vectors involved in the expression.

Accordingly, the elastic strain energy $C_E$ is defined as the potential mechanical energy stored in the configuration of an elastic body as work is performed to distort its volume or shape.

$$\begin{aligned} C_E &= \tfrac{1}{2}\int_V (\boldsymbol{\varepsilon}-\boldsymbol{\varepsilon}^{th})^T \mathbf{D}(\boldsymbol{\varepsilon}-\boldsymbol{\varepsilon}^{th})dV \\ &= \tfrac{1}{2}\int_V \boldsymbol{\varepsilon}^T\mathbf{D}\boldsymbol{\varepsilon}dV - \int_V \boldsymbol{\varepsilon}^T\mathbf{D}\boldsymbol{\varepsilon}^{th}dV + \tfrac{1}{2}\int_V (\boldsymbol{\varepsilon}^{th})^T\mathbf{D}\boldsymbol{\varepsilon}^{th}dV \end{aligned} \quad [4.42]$$

where the third integral term on the right refers to the energy generated by the initial thermal strain. It equals the total strain energy when the domain is fully constrained with $\boldsymbol{\varepsilon} = 0$. Based on the finite element method, the above relation can be discretized as:

$$\begin{aligned} C_E &= \tfrac{1}{2}\mathbf{U}^T\mathbf{K}\mathbf{U} - \mathbf{U}^T\mathbf{F}^{th} + C_E^{ith} \\ C_E^{ith} &= \tfrac{1}{2}\int_\Omega (\boldsymbol{\varepsilon}^{th})^T\mathbf{D}\boldsymbol{\varepsilon}^{th}dV \end{aligned} \quad [4.43]$$

where $C_E^{ith}$ is the initial thermal strain energy. In addition, it can easily be proved that the elastic strain energy consists of two independent parts contributed from the mechanical and thermal loads, respectively.

$$\begin{aligned} C_E &= C_E^m + C_E^{th} \\ C_E^m &= \tfrac{1}{2}(\mathbf{F}^m)^T\mathbf{K}^{-1}\mathbf{F}^m = C^m \\ C_E^{th} &= -\tfrac{1}{2}(\mathbf{F}^{th})^T\mathbf{K}^{-1}\mathbf{F}^{th} + \tfrac{1}{2}\int_V (\boldsymbol{\varepsilon}^{th})^T\mathbf{D}\boldsymbol{\varepsilon}^{th}dV = -C^{th} + C_E^{ith} \end{aligned} \quad [4.44]$$

## 4.5.2. *Comparisons between mean compliance and elastic strain energy*

### 4.5.2.1. *Comparisons in analysis*

It is important to note that the following relationship holds between the elastic strain energy $C_E$ and the mean compliance $C$.

$$C_E = C - \mathbf{U}^T\mathbf{F}^{th} + C_E^{ith} \quad [4.45]$$

Clearly, $C_E$ is equivalent to $C$ when thermal loads disappear with $\Delta T = 0$. This is the common case mostly studied in topology optimization. However, for $\Delta T \neq 0$, $C_E$ and $C$ have different expressions and should be carefully investigated in the formulation of the optimization model.

To some extent, the strain energy measures the stressed state of a structure. Physically, if a structure deforms in a free stress state, the strain energy tends to zero.

**Figure 4.23.** *Illustration of a slender bar under uniform heating: a) free expansion case; b) completely constrained case*

For example, Figure 4.23 illustrates a uniformly heated slender bar only with axial thermal deformation. Suppose the bar is fixed at two ends or only at one end. It therefore follows that

Case (a): $C = 1/2\, EAL\alpha^2 (\Delta T)^2$, $C_E = 0$, $\varepsilon^{m} = 0$, $\varepsilon^{th} = \alpha\Delta T$

Case (b): $C = 0$, $C_E = \frac{1}{2} EAL\alpha^2 (\Delta T)^2 = C_E^{ith}$, $\varepsilon^{m} = -\alpha\Delta T$, $\varepsilon^{th} = \alpha\Delta T$ [4.46]

where, $E$, $A$, $L$ and $\alpha$ denote the Young's modulus, cross-section area, length and coefficient of thermal expansion. It is seen that the mean compliance and strain energy change from one case to the other case. In case (a), no thermal stress exists because the bar is in the state of free expansion. In case (b), the thermal stress is calculated as $\sigma = -\alpha E\Delta T$.

In fact, the von-Mises stress can be expressed in quadratic form of the stress vector,

$$(\sigma^{VM})^2 = \boldsymbol{\sigma}^T \mathbf{J} \boldsymbol{\sigma} \qquad [4.47]$$

where **J** is a constant symmetrical matrix of form similar to the inverse of the elastic matrix $\mathbf{D}^{-1}$. For 2D plane stress problems and 3D problems, **J** is expressed as,

$$\mathbf{J} = \begin{bmatrix} 1 & -1/2 & 0 \\ -1/2 & 1 & 0 \\ 0 & 0 & 3 \end{bmatrix} \quad \text{for 2D problems}$$

$$\mathbf{J} = \begin{bmatrix} \begin{bmatrix} 1 & -1/2 & -1/2 \\ -1/2 & 1 & -1/2 \\ -1/2 & -1/2 & 1 \end{bmatrix} & \mathbf{0}_{3\times3} \\ \mathbf{0}_{3\times3} & 3\begin{bmatrix} 1 & 0 & 0 \\ 0 & 1 & 0 \\ 0 & 0 & 1 \end{bmatrix} \end{bmatrix} \quad \text{for 3D problems} \qquad [4.48]$$

In comparison, $\mathbf{D}^{-1}$ can be written as

$$\mathbf{D}^{-1} = \frac{1}{E}\begin{bmatrix} 1 & -\mu & 0 \\ -\mu & 1 & 0 \\ 0 & 0 & 2(1+\mu) \end{bmatrix} \quad \text{for 2D problems}$$

$$\mathbf{D}^{-1} = \frac{1}{E}\begin{bmatrix} \begin{bmatrix} 1 & -\mu & -\mu \\ -\mu & 1 & -\mu \\ -\mu & -\mu & 1 \end{bmatrix} & \mathbf{0}_{3\times3} \\ \mathbf{0}_{3\times3} & 2(1+\mu)\begin{bmatrix} 1 & 0 & 0 \\ 0 & 1 & 0 \\ 0 & 0 & 1 \end{bmatrix} \end{bmatrix} \quad \text{for 3D problems} \qquad [4.49]$$

This indicates that the element strain energy is close to element von-Mises stress in form. The reduction of maximum value of the element strain energy can result in the reduction of maximum value of element von-Mises stress. Therefore, the strain energy of an element is related to the stress state to some extent.

### 4.5.2.2. *Comparisons in topology optimization*

In this chapter, two formulations are investigated for topology optimization of thermo-elastic problems subjected to the volume constraint. One is to use the mean compliance as the objective function. The other is to use the elastic strain energy as the objective function.

The aim of this section is to show how the mean compliance, strain energy and von-Mises stress change when each formulation is used for topology optimization. To avoid the local stress concentration at the load application point, distributed loads or point-wise loads applied on a non-designable domain are considered as shown in Figure 4.24.

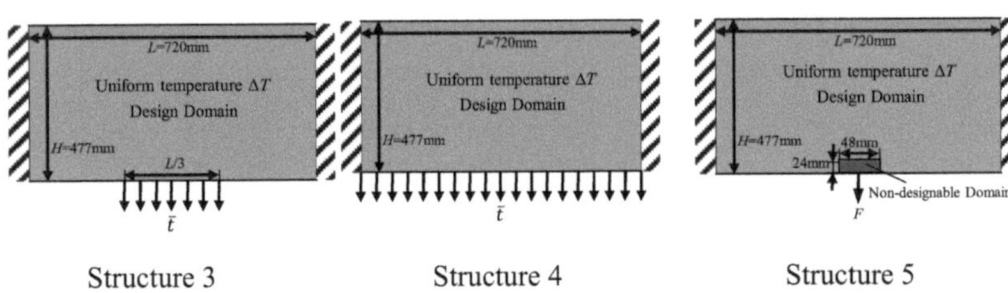


**Figure 4.24.** *Biclamped domain under thermo-mechanical loads*

In these three tests, assume that the Young's modulus and coefficient of thermal expansion of the solid material are $E = 210$ GPa and $\alpha = 1.1\times10^{-5}/°\text{C}$, respectively. The domain is uniformly heated with a temperature variation $\Delta T = 50$ K. The distributed load is 100 N/m and the point load is 10 kN. The whole domain is discretized with a 60 × 40 mesh of eight-node bilinear elements for finite element analysis. Element size is about 12 mm×11.9 mm. The volume fraction 40% is used as the upper bound of the volume constraint.

Figures 4.25 and 4.26 show the optimized configurations and corresponding distributions of von-Mises stresses for the mean compliance and strain energy minimizations, respectively. It should be remarked that for elements with intermediate pseudo-densities (not 0/1), stress computations become dependent on the choice of the interpolation parameters. This is, however, negligible if few intermediate elements exist in the final configuration. For the purpose of illustration, Figure 4.27 gives the iteration histories of structure 3. Based on numerical results, the following conclusions can be drawn out.

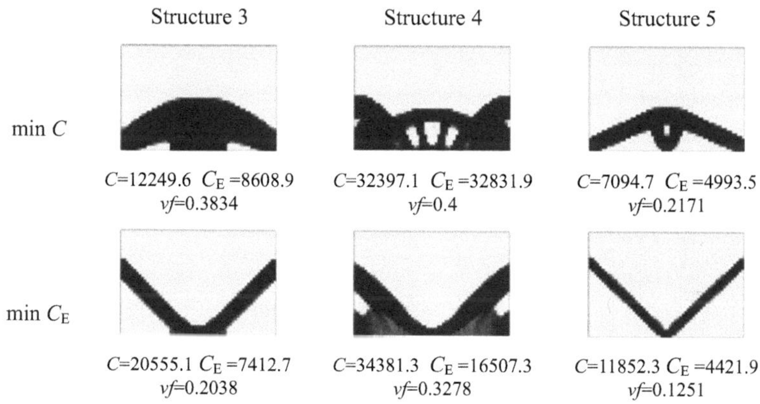


**Figure 4.25.** *Comparisons of optimized configurations*

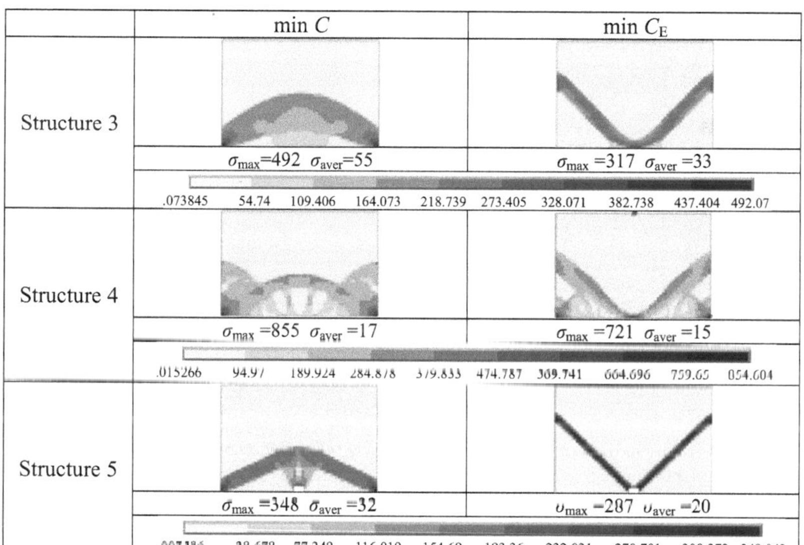


**Figure 4.26.** *Stress distributions of the optimized configurations*

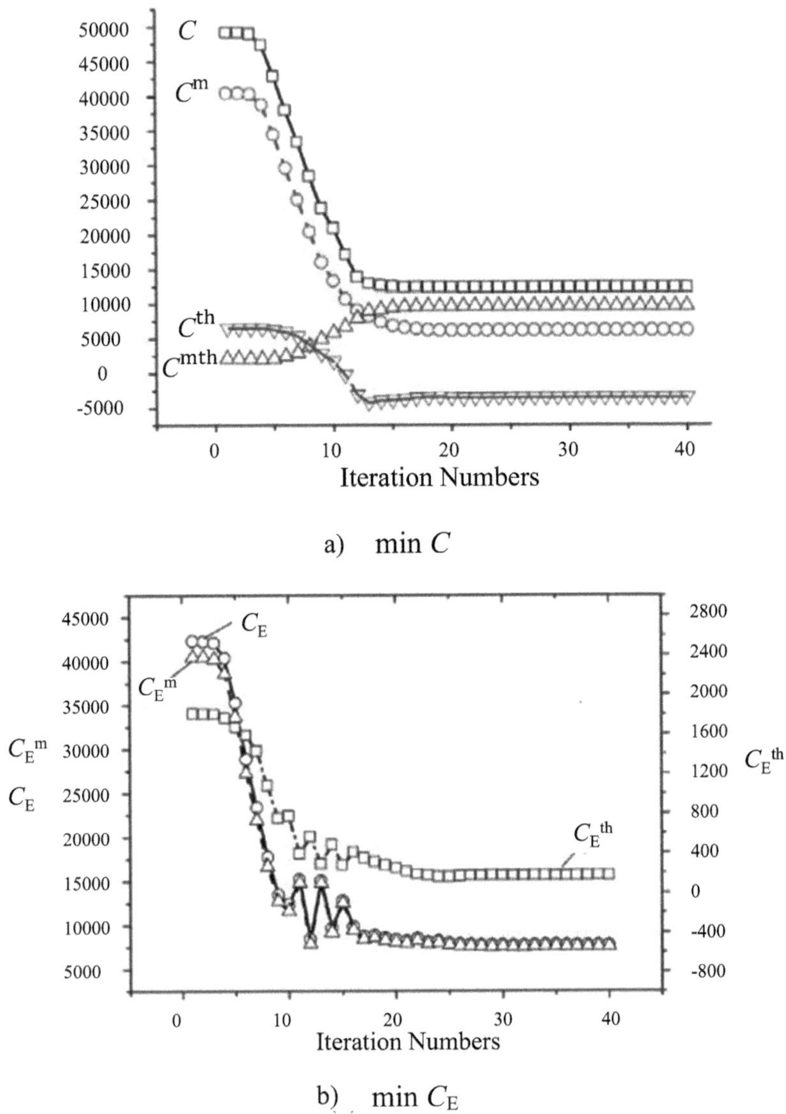


**Figure 4.27.** *Iteration histories (structure 3). For a color version of this figure, see www.iste.co.uk/zhang/topology.zip*

1) The mean compliance minimization can achieve a lower mean compliance value than that using the strain energy minimization.

Likewise, the strain energy minimization can produce a lower strain energy value than that using the mean compliance minimization.

2) As mentioned previously, the strain energy depends upon the stressed state of the structure. So the strain energy minimization tends to release the constraints related to the thermal deformation of the structure as much as possible. This is why less material is used and retained in both fixed ends of the domain after optimization. More importantly, the maximum and average stress values in the optimized configuration using the strain energy minimization are reduced and more uniformly distributed than those in the mean compliance minimization. In other words, the strain energy minimization provides a better strength design than the mean compliance minimization.

3) Since $C^{\mathrm{mth}}$ is the coupling term of mechanical and thermal loads in the mean compliance, $C^{\mathrm{mth}}$ even decreases to negative values along with the increase in $C^{\mathrm{th}}$ and the decrease in $C^{\mathrm{m}}$ in the mean compliance minimization. Instead, $C_{\mathrm{E}}{}^{\mathrm{m}}$ and $C_{\mathrm{E}}{}^{\mathrm{th}}$ are two independent terms. In the strain energy minimization, both $C_{\mathrm{E}}{}^{\mathrm{m}}$ and$C_{\mathrm{E}}{}^{\mathrm{th}}$ decrease. As a result, the value of $C_{\mathrm{E}}{}^{\mathrm{m}}$ becomes greatly dominant over $C_{\mathrm{E}}{}^{\mathrm{th}}$ in the composition of $C_{\mathrm{E}}$.

### 4.5.3. *Effects of thermal and mechanical loads on the optimized configurations*

As shown in numerical tests above, the optimized configuration may change greatly depending upon whether the load is dominated by the mechanical or thermal load. In order to reveal the effects of each load upon the structural compliance and strain energy, thermal and mechanical load sensitivity analyses are carried out. To do this, load coefficients $\lambda^{\mathrm{m}}$ and $\lambda^{\mathrm{th}}$ are first introduced as coefficients applied to initial mechanical load vector $\mathbf{F}^{\mathrm{m0}}$ and thermal load vector $\mathbf{F}^{\mathrm{th0}}$.

$$\begin{aligned} &\mathbf{F}^{\mathrm{m}} = \lambda^{\mathrm{m}}\mathbf{F}^{\mathrm{m0}}, \\ &\mathbf{F}^{\mathrm{th}} = \lambda^{\mathrm{th}}\mathbf{F}^{\mathrm{th0}}, \quad \Delta T = \lambda^{\mathrm{th}}\Delta T_0 \end{aligned} \qquad [4.50]$$

Consider now the objective function defined by the mean compliance or the strain energy. The following expressions can be derived

$$\begin{aligned} C &= C^{\mathrm{m}} + C^{\mathrm{th}} + C^{\mathrm{mth}} = (\lambda^{\mathrm{m}})^2 C^{\mathrm{m0}} + (\lambda^{\mathrm{th}})^2 C^{\mathrm{th0}} + \lambda^{\mathrm{m}} \lambda^{\mathrm{th}} C^{\mathrm{mth0}} \\ C_{\mathrm{E}} &= C_{\mathrm{E}}^{\mathrm{m}} + C_{\mathrm{E}}^{\mathrm{th}} = (\lambda^{\mathrm{m}})^2 C_{\mathrm{E}}^{\mathrm{m0}} + (\lambda^{\mathrm{th}})^2 C_{\mathrm{E}}^{\mathrm{th0}} \end{aligned} \qquad [4.51]$$

where $C^{\mathrm{m0}}$, $C^{\mathrm{th0}}$, $C^{\mathrm{mth0}}$, $C_{\mathrm{E}}^{\mathrm{m0}}$ and $C_{\mathrm{E}}^{\mathrm{th0}}$ refer to corresponding parts of the mean compliance and the strain energy under initial loads.

Without loss of generality, the topology optimization problem can be written in the following general form

$$\begin{array}{lll} \text{find} & \boldsymbol{\eta} = \{\eta_i\} & i = 1,2,\ldots,n_{\mathrm{e}} \\ \min & f(\boldsymbol{\eta},\boldsymbol{\lambda}) & \boldsymbol{\lambda} = (\lambda_1,\lambda_2,\cdots,\lambda_s) \\ \text{s.t} & h(\boldsymbol{\eta}) \le 1 & \end{array} \qquad [4.52]$$

where $\boldsymbol{\eta}$ denotes the $n_{\mathrm{e}}$-dimensional vector of topological design variables; $\boldsymbol{\lambda}$ denotes the $s$-dimensional vector of fixed parameters, i.e. load coefficients in our case; $f$ denotes the objective function; $h$ denotes the volume constraint function, which is independent of fixed parameters.

Suppose the gradient of the objective function $f$ with respect to $\boldsymbol{\eta}$ is expressed as

$$\nabla \mathrm{f}(\boldsymbol{\eta},\boldsymbol{\lambda}) = \left[ \frac{\partial \mathrm{f}(\boldsymbol{\eta},\boldsymbol{\lambda})}{\partial \eta_1} \; \frac{\partial \mathrm{f}(\boldsymbol{\eta},\boldsymbol{\lambda})}{\partial \eta_2} \; \cdots \; \frac{\partial \mathrm{f}(\boldsymbol{\eta},\boldsymbol{\lambda})}{\partial \eta_{n_{\mathrm{e}}}} \right]^{\mathrm{T}} \qquad [4.53]$$

The Karush–Kuhn–Tucker (KKT) condition can be used to characterize the influences of load amplitudes characterized by load coefficients $\lambda^{\mathrm{m}}$ and $\lambda^{\mathrm{th}}$ upon the optimum solution related to each metric. In our case, it is theoretically stated that at an optimum point, the negative gradient of the objective function can be expressed as a linear combination of gradients of active constraints with non-negative Lagrangian multipliers. Since the gradients of the volume constraint and side constraints involved in the topology optimization problem are constant and independent of $\lambda^{m}$ and $\lambda^{th}$ in topology optimization, the deviation of the optimized solution will be made possible only if any perturbation of $\lambda^{m}$ and $\lambda^{th}$ changes the gradient direction of the objective function. Hence, the gradient deviation of the objective function can be quantified by the angle between the perturbed gradient vector with $\Delta\lambda$ and the original gradient vector to measure the gradient direction variation.

$$\cos\theta = \frac{\nabla f(\boldsymbol{\eta}, \boldsymbol{\lambda} + \Delta\boldsymbol{\lambda}) \cdot \nabla f(\boldsymbol{\eta}, \boldsymbol{\lambda})}{\left\|\nabla f(\boldsymbol{\eta}, \boldsymbol{\lambda} + \Delta\boldsymbol{\lambda})\right\| \cdot \left\|\nabla f(\boldsymbol{\eta}, \boldsymbol{\lambda})\right\|} \quad [4.54]$$

$\cos\theta$ is an important indicator of the load effects upon the optimized configuration. If the value of $\cos\theta$ is close to 1, it means that the gradient directions before and after the load change are nearly parallel so that small changes will appear in the optimized configurations. Otherwise, closer to 0 for $\cos\theta$, greater changes will appear in the optimized configurations.

The structure 1 is tested here for detailed illustrations. Suppose that the initial mechanical load is $F^{m0} = 10^4$ N and temperature variation is $\Delta T_0 = 1$ K. Figure 4.28 shows the corresponding gradient direction variations of each metric when only thermal loads increase with respect to the reference value of $\lambda^{th} = -5$. Topologies of the optimized structures achieved by both formulations are quite different at $\lambda^{th} \neq 0$. Notice that at $\Delta T = \lambda^{th} = 0$, the mean compliance is identical to the strain energy for the definition of objective function.

Figure 4.28 indicates the significant changes of optimized configurations compared with the reference when $\cos\theta$ increases and decreases. As shown in Figure 4.29, the same conclusion can be drawn out when mechanical loads change.

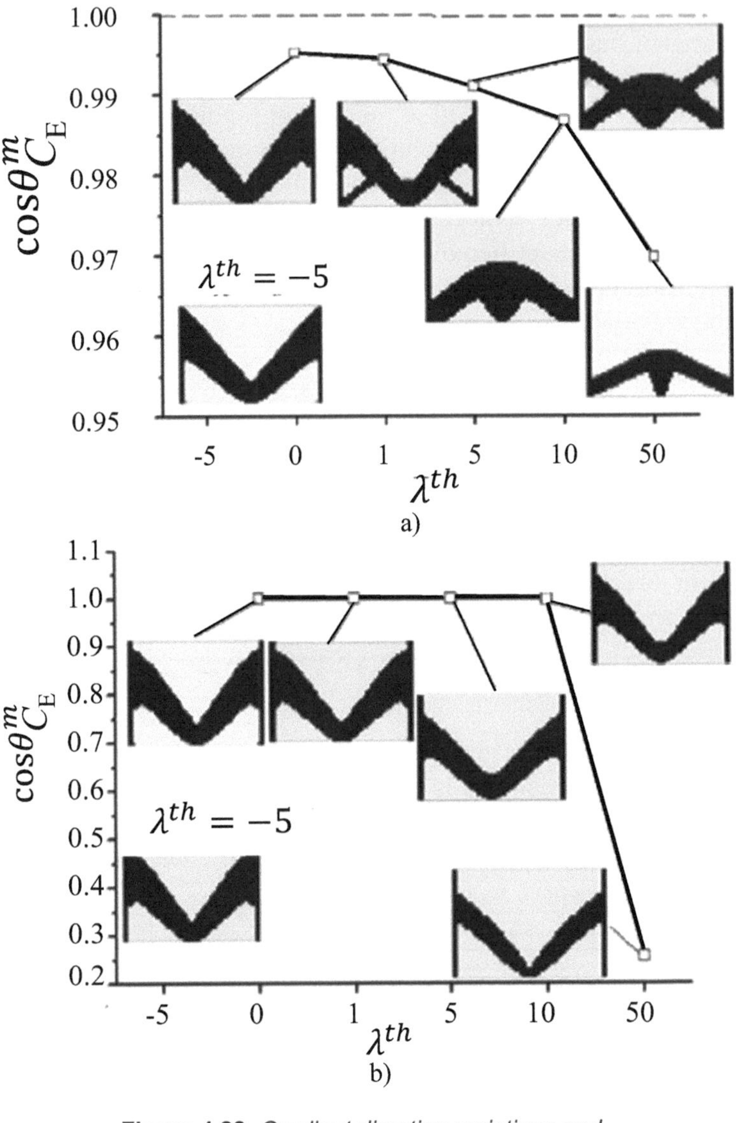


**Figure 4.28.** *Gradient direction variations and corresponding optimized configurations under different thermal loads ($\lambda^{th}$ = –5 as the reference value): a) mean compliance minimization; b) strain energy minimization*

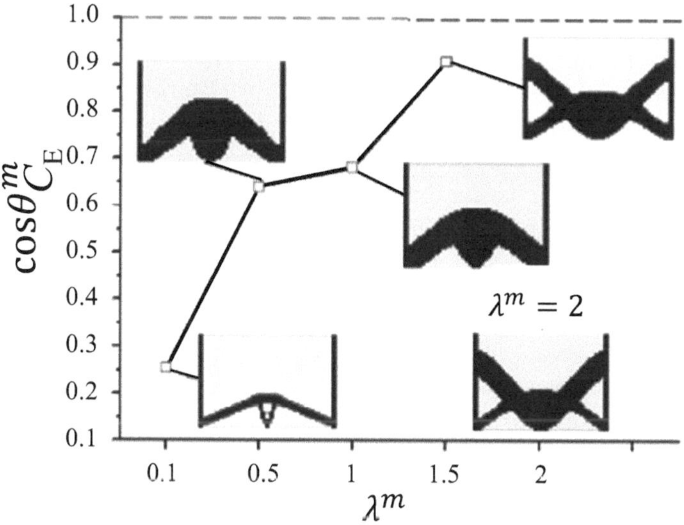


a)

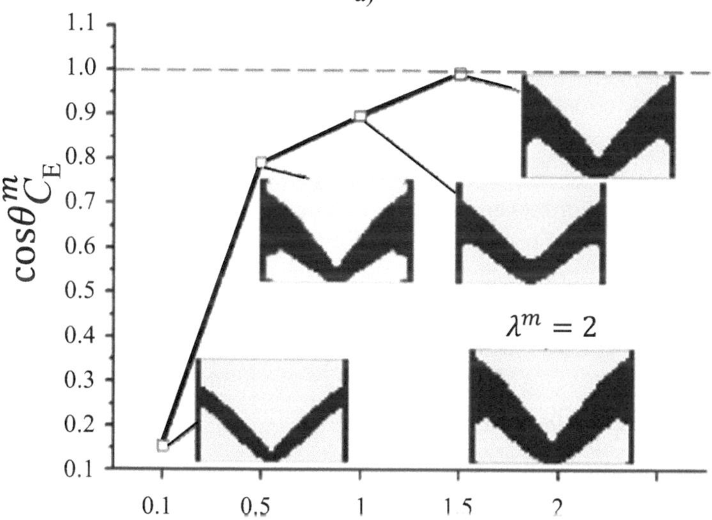


b)

**Figure 4.29.** *Gradient direction variations and corresponding optimized configurations under different mechanical loads ($\lambda^m$ = 2 as the reference value): a) mean compliance minimization; b) strain energy minimization*

Although the optimized configurations related to both formulations are sensitive to the variations of thermal and mechanical loads, the influence measured by the load sensitivity is physically different. In the mean compliance minimization, the load change will change the pattern of material layout. In the strain energy minimization, the load change will only change the member thickness, while the pattern of material layout remains unchanged.

## 4.6. Conclusions

In this chapter, the concept of TSC is first introduced as an inherent property of material to favor the formulation of topology optimization problems involving thermo-elastic stress. With numerical examples, the RAMP model demonstrates superior advantages over the SIMP model for this kind of problem. Moreover, the internal RAMP parameters are studied to reveal the penalty effects upon the structural stiffness and thermal stress load.

A mass constraint is implemented into thermo-elastic topology optimization for structures of multiphase materials to reflect the engineering design needs of light weightness. To avoid the possible mixed material status within each element, the optimization formulation is improved by introducing the artificial penalty term at the objective function level. The influence of the additive parameter $\tilde{a}$ is studied and the variable $\tilde{a}$ is then suggested. Due to the presence of thermal stress loads, numerical tests highlight that non-uniform initial weighting in the UMMI might be more advantageous than the uniform initial weighting in obtaining the optimum solution. In contrast, for the case with pure mechanical loads, uniform initial weighting is preferable to avoid the local optimum.

Two formulations related to the mean compliance minimization and strain energy minimization are studied for thermo-elastic topology optimization. Based on theoretical and numerical comparisons, it is found that both formulations lead to different optimized configurations when thermo-elastic load exists. Meanwhile, the strain energy minimization is more beneficial for the stress reduction than the mean compliance minimization. In addition, although the optimized configurations related to both formulations are sensitive to the variations of thermal and mechanical loads, the influence measured by the load sensitivity is physically different. In the mean compliance minimization, the load change will change the pattern of material layout. In the strain energy minimization, the load change will only change the member thickness, while the pattern of material layout remains unchanged.

# 5

# Integrated Layout and Topology Optimization

## 5.1. Introduction to Integrated optimization

A multicomponent system is a structural system consisting of a certain number of components of specific form, a container and the supporting structure that interconnects the components and the container for its integrity. Most structural systems in mechanical and aerospace engineering can be considered as a kind of multicomponent system, as shown in Figure 5.1. Compactness, structural efficiency, static and dynamic responses have to be optimized for the functionality and mechanical performance. On the one hand, the given components are assembled in the limited space of the container to satisfy various functional constraints, which are a packing optimization or a layout optimization. On the other hand, proper supporting structures have to be identified to satisfy the mechanical performances of the system, which is a typical topology optimization. In this work, the integrated layout and topology optimization are discussed for the maximum rigidity design where the spatial placement of components and the configuration of the supporting structure have to be optimized simultaneously, as illustrated in Figure 5.2.

Generally speaking, there are two main difficulties involved in the integrated optimization design [ZHU 06b]. First, non-overlapping constraints will be properly defined to avoid the collision of components. Second, the integrated finite element model including the modeling of the components, structures and their interfaces will be studied to favor the simultaneous design.

**Figure 5.1.** *Typical multicomponent system*

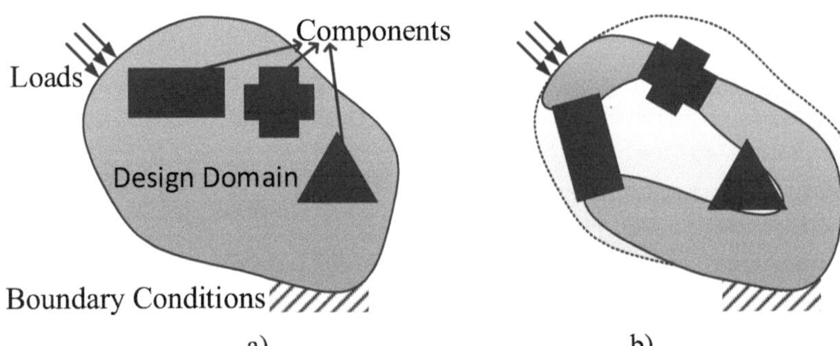


**Figure 5.2.** *Illustration of integrated layout and topology optimization design of multicomponent systems: a) initial definition; b) optimized design*

## 5.2. Finite-circle method

### 5.2.1. *Formulation of finite-circle method*

According to the definition of integrated optimization, packing optimization with non-overlapping constraints has to be performed to figure out the proper layout of components. Until now, packing optimization problems are still CAD based. Simple geometrical or physical measures like the system compactness, gravity center, configuration cost, etc., are optimized in the packing design by assigning the location and orientation of the components as design variables. As indicated by Cagan *et al.* [CAG 98, CAG 02], the underlying difficulties are concerned with the modeling of the components, the packing area and selection of efficient search strategies, etc. Among others, one of the key issues is that the non-overlapping constraints have to be properly introduced in order to avoid both the components' overlapping and their overlapping with the design domain boundaries. Theoretically, it was proved that packing optimization is a type of NP-hard problem [DEB 88]. Varieties of component shapes and design domain boundaries will lead to high nonlinearity and even discontinuity of the non-overlap constraint functions. Consequently, the gradient-based optimization algorithms will be strongly limited in solving packing problems when the components or the design domain boundary have complex and concave shapes. Even for the simplest bin-packing problem with rectangular components, different overlap cases exist and should be analyzed carefully [PAL 06, HUA 07].

Practically, techniques such as octrees [MEA 82, SAM 89], sphere trees [MOO 02, HUB 93, QUI 94] and S-bounds based trees [CAM 91] have been proposed to detect object collision. These techniques approximate the components with various levels of small cubes or spheres and refine the model partition iteratively. However, these methods are limited to detecting rather than evaluating the overlapping. More interesting information cannot be provided to identify the search directions and to verify attainment of the optimum solution. For more overlapping detection methods, we can refer to the work of Lin and Gottschalk [LIN 98].

In this section, the Finite-circle method (FCM) is presented to adapt gradient-based algorithms to the packing design. Consider a 2D packing problem as shown in Figure 5.3. Suppose several components will be located inside the design domain and no overlapping will be found between different components. Mathematically, the following conditions should be retained

$$\begin{cases} \forall \psi = 1,2,\ldots,n_c; \\ \text{s.t.: } \mathbf{\Omega}_\psi(x_\psi, y_\psi, \theta_\psi) \subset \mathbf{\Omega}_D \\ \forall \psi 1 = 1,2,\ldots,n_c;\ \psi 2 = 1,2,\ldots,n_c;\ \psi 1 \neq \psi 2 \\ \text{s.t.: } \mathbf{\Omega}_{\psi 1}(x_{\psi 1}, y_{\psi 1}, \theta_{\psi 1}) \cap \mathbf{\Omega}_{\psi 2}(x_{\psi 2}, y_{\psi 2}, \theta_{\psi 2}) = \varnothing \end{cases} \qquad [5.1]$$

where $\mathbf{\Omega}_\psi$ and $\mathbf{\Omega}_D$ denote the area occupied by the $\psi$th component and the global design domain, respectively. $\mathbf{\Omega}_\psi$is described as the function of the location and orientation of the component, i.e. $(x_\psi, y_\psi, \theta_\psi)$. $n_c$ is the number of components. $\psi 1$ and $\psi 2$ denote two different components. This equation is a symbolic presentation. When gradient-based algorithms are applied, it is necessary to have the design sensitivities in order to understand how much the components are overlapped with each other, in which direction they should move to escape from the overlapping and possible overlapping.

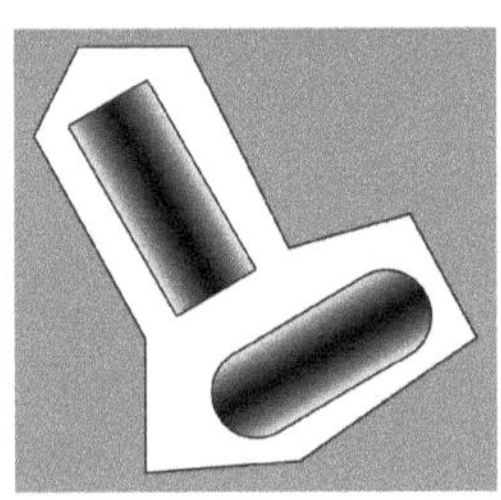

**Figure 5.3.** *A 2D packing problem*

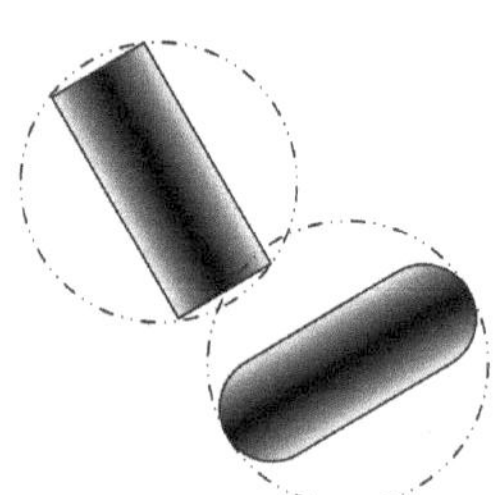

**Figure 5.4.** *Components approximation with single circle*

Although it is difficult to describe overlapping for components with arbitrary shapes, overlapping between circles and spheres can be easily calculated by comparing the distances of the centers with the summation of their radii. This is the idea of approximating all the components of arbitrary shapes with circles or spheres. For clarity of presentation, only the formulation related to 2D components is discussed here. If only one single circle is used for the approximation of one component as shown in Figure 5.4, the approximation error will be large. Consequently, the two components are still far away from each other but the circles have already overlapped. Here, FCM uses the idea similar to the sphere-trees. The components and the design domain are approximately modeled with numbers of circles (2D) or spheres (3D) as shown in Figure 5.5. For each component, a family of circles may have different radii and be placed at different locations to approximately cover the boundary of the components. Clearly, the approximation accuracy can be improved by refining the circles' definition or simply using more circles.

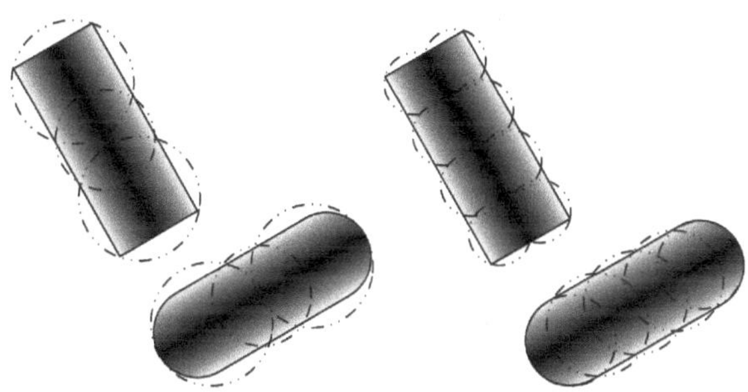

**Figure 5.5.** *The FCM approximation of the components*

In this way, the complex non-overlapping constraint can be transformed and simplified into a standard form of constraint between circles

$$\left\| O_{\psi 1,\varsigma 1} O_{\psi 2,\varsigma 2} \right\| \geq R_{\psi 1,\varsigma 1} + R_{\psi 2,\varsigma 2} \qquad [5.2]$$

where $O_{\psi 1,\varsigma 1}$ is the center of the ($\varsigma 1$)th circle defined in the ($\psi 1$)th component. $R_{\psi 1,\varsigma 1}$ is the corresponding radius. This relationship refers to the distance condition between the circle centers. The gradients with respect to the involved design variables can thus be easily calculated by differentiating the constraint function. From this viewpoint, FCM can be considered a common approach for packing optimization problems with components and design domain of arbitrary shapes.

Compared with the sphere-trees methods, the iterative approximation refinements of the circles' discretization are not employed in FCM, which implies that the FCM uses a fixed number of design constraints during the packing optimization. In fact, FCM is more than a collision detection method. It identifies how to relocate the components to avoid overlap by calculating the sensitivities of the distances between the circles with respect to the location and orientation of the components. Note that although FCM proposes to favor the gradient-based algorithms, it does not limit the application of the gradient-free methods.

A simple packing optimization problem consisting of six identical equilateral triangle components and an equilateral hexagon design domain is illustrated in Figure 5.6. The characteristic length $l$ is now assigned as the edge length of the triangular component and each component is approximated with nine identical circles with a maximum approximation error of 7.22%$l$. Likewise, the design domain is approximated with six big circles and the maximum error is 3%$l$. The problem is now to find the optimal locations of all components inside the design domain with a minimum packing area. Mathematically, the problem can be stated as the height minimization of the equilateral hexagon.

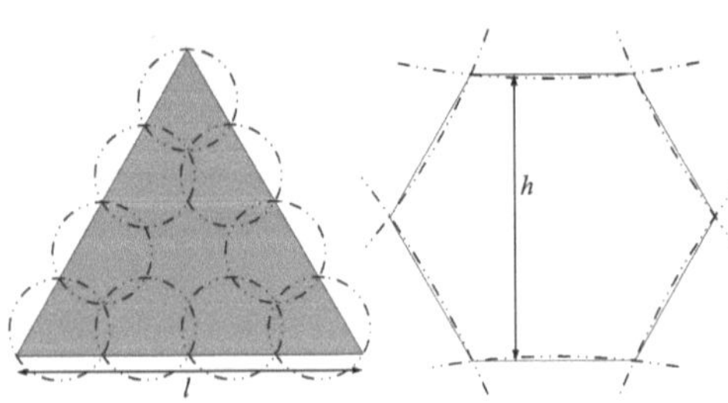


**Figure 5.6.** *The triangle component and the design domain*

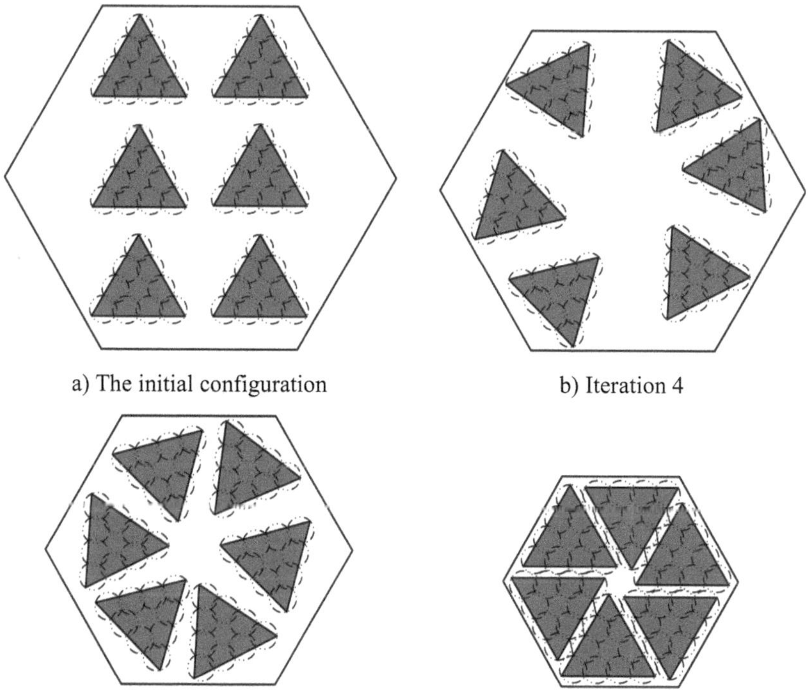

a) The initial configuration

b) Iteration 4

c) Iteration 8

d) The final configuration

**Figure 5.7.** *The design iteration and the optimized result*

Finally, 1,215 constraints are retained to avoid overlap between the components and 324 constraints to keep all the components inside the design domain. Based on the initial configuration shown in Figure 5.7(a), the optimization problem is programmed and sensitivity analysis is carried out in the framework of Boss-Quattro, the design process converges to the configuration shown in Figure 5.7(d) after 13 iterations. Because FCM uses conservative approximation, there are always safety gaps between the design domain contour and the components.

To improve the approximation accuracy, more circles are used as shown in Figure 5.8 with the uniform discretization. Although the approximation error is reduced, the great number of the circles leads to a large number of design constrains which requires tremendous computing time.

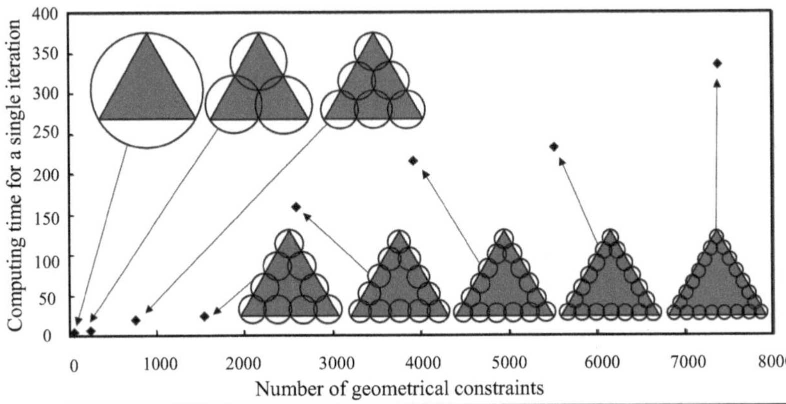


| Number of circles | 1 | 3 | 6 | 9 | 12 | 15 | 18 | 21 |
|---|---|---|---|---|---|---|---|---|
| Number of constraints | 51 | 243 | 756 | 1539 | 2592 | 3915 | 5508 | 7371 |
| Computing time for one iteration (s) | 4.90 | 6.68 | 19.48 | 24.71 | 158.79 | 216.46 | 232.82 | 335.28 |

**Figure 5.8.** *Effects of circle and constraint numbers upon the computing time for one single iteration*

## 5.2.2. *Improved adaptive constraint aggregation*

FCM possesses the advantage of simple and explicit formulation which can be easily differentiated to yield sensitivities for gradient-based optimizations. Moreover, FCM can easily be extended to deal with 3D overlapping problems by using spheres instead of circles.

However, precise description of components' geometries requires more circles which will lead to large numbers of overlapping constraints and costly computing times when the packing optimization problem is solved iteratively. It is therefore crucial to use fewer circles on the one hand while maintaining the approximation precision. On the other hand, some constraint aggregation approaches such as KS function, etc., are applicable to reduce the number of constraints. The standard form of KS function is expressed as:

$$KS(\tau) = \frac{1}{\tau}\ln\sum_{m=1}^{N_g}\exp(\tau g_m) \tag{5.3}$$

where $N_g$ stands for the number of constraints. $\tau$ is the aggregation parameter. Generally, the constraints should be normalized into the same scale, which will be helpful, especially when the constraints are in different magnitudes.

Normally, an equivalent form is used to avoid some numerical difficulties.

$$KS(\tau) = g_{\max} + \frac{1}{\tau}\ln\sum_{m=1}^{N_g}\exp\left[\tau\left(g_m - g_{m\max}\right)\right] \tag{5.4}$$

where $g_{\max}$ stands for the maximum value of all the constraints. We also have

$$\begin{aligned} & g_{\max} < KS(\tau) < g_{\max} + \frac{1}{\tau}\ln\left(N_g\right) \\ & \lim_{\mu\to+\infty} KS(\tau) = g_{\max} \end{aligned} \tag{5.5}$$

Obviously, choosing larger values for aggregation parameters has the advantages of a precise description of the feasible region. However, too large a value of the aggregation parameter will lead to an ill-conditioned Hessian matrix of the KS function and result in optimization instability when the current design point is located very close to, or directly at the points where multiple constraints are active. Practically, $\tau$ should be properly defined considering both the precision and stability of optimization iteration.

In the existing adaptive aggregation method [MAR 05, POO 07], $|\mathrm{d}KS/\mathrm{d}\tau|$ is chosen as the criterion to determine the aggregation parameters adaptively during each iteration.

$$\left|\frac{\mathrm{d}KS}{\mathrm{d}\tau}\right| \le \delta \tag{5.6}$$

$\delta$ is a prescribed small positive value. By assuming a linear relationship between $|\mathrm{d}KS/\mathrm{d}\tau|$ and $\tau$ in logarithmic scale, the derivative of $|\mathrm{d}KS/\mathrm{d}\tau|$ w.r.t. $\tau$ was then obtained with a finite difference scheme. The desired value of $\tau = \tau_d$ satisfying equation [5.6] was thus directly obtained by

$$\frac{\log\left|\frac{\mathrm{d}KS}{\mathrm{d}\tau}\Big|_{\tau=\tau_1}\right| - \log\left|\frac{\mathrm{d}KS}{\mathrm{d}\tau}\Big|_{\tau=\tau_0}\right|}{\log\tau_1 - \log\tau_0} = \frac{\log\left|\frac{\mathrm{d}KS}{\mathrm{d}\tau}\Big|_{\tau=\tau_d}\right| - \log\left|\frac{\mathrm{d}KS}{\mathrm{d}\tau}\Big|_{\tau=\tau_0}\right|}{\log\tau_d - \log\tau_0} \quad [5.7]$$

where $\tau_0$ is assigned as an initial value of 50, and $\tau_1 = \tau_0 + 10^{-3}$. Note that when $\tau = \tau_d$, $|\mathrm{d}KS/\mathrm{d}\tau|$ is equal to $\delta$.

Here, we explain the conditions in which equation [5.7] holds. By considering the standard form of the KS function, we have its first derivative

$$\frac{\mathrm{d}KS}{\mathrm{d}\tau} = \frac{\sum_{m=1}^{Ng}(g_m - KS)\exp(\tau g_m)}{\tau\sum_{m=1}^{Ng}\exp(\tau g_m)} \quad [5.8]$$

In most cases, the linear relationship between $|\mathrm{d}KS/\mathrm{d}\tau|$ and $\tau$ in logarithmic scale cannot be attained. However, in some special situations, e.g. when all the constraints' values are equal, the aggregation absolute error can be written as

$$KS - g_{\max} = \frac{1}{\tau}\ln N_g \quad [5.9]$$

we have

$$\frac{\mathrm{d}KS}{\mathrm{d}\tau} = -\frac{\ln N_g}{\tau^2} \quad [5.10]$$

Therefore, the linear relationship between $|\mathrm{d}KS/\mathrm{d}\tau|$ and $\tau$ in logarithmic scale can be obtained.

$$\log\left|\frac{\mathrm{d}KS}{\mathrm{d}\tau}\right| = \log(\ln N_g) - 2\log\tau \quad [5.11]$$

As the linear assumption only exists in few particular cases, the adaptive approach may therefore fail in obtaining proper aggregation parameters. To solve this problem, an improved adaptive approach based on the Steffensen iteration is presented. First, we have

$$\frac{\mathrm{d}^2KS}{\mathrm{d}\tau^2} = \frac{2a(a\ln a - \tau b) + \tau^2(ac - b^2)}{\tau^3 a^2} \quad [5.12]$$

where

$$\begin{aligned} a &= \sum_{m=1}^{N_g} \exp(\tau g_m); \\ b &= \sum_{m=1}^{N_g} g_m \exp(\tau g_m); \\ c &= \sum_{m=1}^{N_g} g_m^2 \exp(\tau g_m). \end{aligned} \quad [5.13]$$

It is easy to obtain

$$\begin{aligned} &a > 0; \\ &a\ln a - \tau b = a\tau KS - \tau b = \tau\sum_{m=1}^{N_g}(KS - g_m)\exp(\tau g_m) > 0; \\ &ac - b^2 = \frac{1}{2}\sum_{m=1}^{N_g}\sum_{n=1}^{N_g}(g_m - g_n)^2\exp(\tau g_m + \tau g_n) \geq 0, \mathrm{m} \neq \mathrm{n}. \end{aligned} \quad [5.14]$$

Consequently, we can prove

$$\frac{\mathrm{d}^2KS}{\mathrm{d}\tau^2} > 0$$

$$\lim_{\mu \to +\infty} \frac{\mathrm{d}KS}{\mathrm{d}\tau} = 0^{-} \quad [5.15]$$

By giving a small positive value $\delta$, there is one and only one positive $\tau_\delta$ satisfying the following equation

$$\left|\mathrm{d}KS/\mathrm{d}\tau\right|_{\tau=\tau_d} = \delta \quad [5.16]$$

So, we propose using an iterative adaptive approach to search the aggregation parameter by introducing both upper and lower bounds for $|\mathrm{d}KS/\mathrm{d}\tau|$.

$$\upsilon\delta \le \left|\mathrm{d}KS/\mathrm{d}\tau\right| \le \delta,\ 0 \le \upsilon \le 1 \quad [5.17]$$

Here, we use the equivalent form of KS function, its derivative can be expressed as

$$\frac{\mathrm{d}KS}{\mathrm{d}\tau} = -\frac{1}{\tau}\left(KS - g_{m\max}\right) + \frac{\sum_{m=1}^{Ng}\left(g_m - g_{m\max}\right)\exp\left[\tau\left(g_m - g_{m\max}\right)\right]}{\tau\sum_{m=1}^{Ng}\exp\left[\tau\left(g_m - g_{m\max}\right)\right]} \quad [5.18]$$

Then, we assume

$$f = \frac{\sum_{m=1}^{Ng}\tau\left(g_m - g_{m\max}\right)\exp\left[\tau\left(g_m - g_{m\max}\right)\right]}{\sum_{m=1}^{Ng}\exp\left[\tau\left(g_m - g_{m\max}\right)\right]} \quad [5.19]$$

which leads to

$$\frac{\mathrm{d}KS}{\mathrm{d}\tau} = -\frac{1}{\tau}\left(KS - g_{m\max}\right) + \frac{f}{\tau^2} \quad [5.20]$$

We use the Steffensen iteration to obtain the root of the following equation

$$-\frac{1}{\tau}(KS - g_{m\max}) + \frac{f}{\tau^2} = \delta \qquad [5.21]$$

The stable point formula chosen for the Steffensen iteration can be reorganized as

$$\tau = \sqrt{\frac{\tau}{\delta}(KS - g_{m\max}) - \frac{f}{\delta}} \qquad [5.22]$$

Then the final formula of the Steffensen iteration can be expressed as

$$\begin{aligned} \varphi(\tau_k) &= \sqrt{\frac{\tau_k}{\delta}(KS_k - g_{m\max}) - \frac{f_k}{\delta}} \\ \tau_{k+1} &= \tau_k - \frac{\left[\varphi(\tau_k) - \tau_k\right]^2}{\varphi\left[\varphi(\tau_k)\right] - 2\varphi(\tau_k) + \tau_k} \end{aligned} \qquad [5.23]$$

where $\tau_k$ is initially set to be 50, a common value chosen in many previous works. Then d$KS$/d$\tau$ is iteratively calculated and evaluated until equation [5.17] is satisfied. Since the Steffensen iteration proves to be at least quadratically convergent, the proper value of $\tau$ can be obtained efficiently.

As a substantial member of non-overlapping constraints have been aggregated into one with high nonlinearity, precise sensitivity analysis becomes an important issue. The finite difference and the analytical scheme are two usual methods to achieve design sensitivities.

The finite difference scheme is easy to implement but leads to inaccurate sensitivities. It has been proven that each design variable in different iterations needs adaptive step lengths to ensure a stable sensitivity output. Typically, tiny finite difference step lengths will not improve the precision as expected due to the numerical calculation errors.

On the contrary, the analytical method holds the advantage of accuracy. However, compared with the finite difference scheme, the analytical method for the aggregated non-overlapping constraint needs more operations due to the tedious formulations and derivations.

As a result, we chose to use the complex step derivative approximation [SQU 98, MAR 03, LAI 08] here. Consider a complex function with one complex variable. It can be expanded by using a Taylor series at the current design point

$$f(x+ih)=f(x)+ihf'(x)-h^2\frac{f''(x)}{2!}-ih^3\frac{f'''(x)}{3!}+\cdots \qquad [5.24]$$

Then, only the imaginary parts are extracted

$$f'(x)=\frac{\mathrm{Im}[f(x+ih)]}{h}+h^2\frac{f'''(x)}{3!}+\cdots \qquad [5.25]$$

Ignoring the high-order items, we obtain an approximation form by using a small value of $h$

$$f'(x)\approx\frac{\mathrm{Im}[f(x+ih)]}{h} \qquad [5.26]$$

This formulation ensures a more accurate sensitivity output than the standard finite difference method. On the one hand, it is based on an $O(h^2)$ approximation of $f'(x)$. On the other hand, subtraction operation leading to roundness error is avoided in the complex formulation.

The complex step derivative approximation used in our problem is expressed as

$$\frac{\partial KS(s_\psi)}{\partial s_\psi}\approx\frac{\mathrm{Im}[KS(s_\psi+ih)]}{h} \qquad [5.27]$$

where Im[KS($s_\psi$+$ih$)] denotes the imaginary output after a small imaginary perturbation $ih$ is brought into the KS function. Normally, we use $h = 10^{-12}$ in our work. Here, $s_\psi$ is one of the geometry design variables such as $x_\psi$, $y_\psi$ or $\theta_\psi$.

## 5.3. Density points and embedded meshing

### 5.3.1. *Definition of the density points*

Another key difficulty of the integrated optimization lies in the description of the interfaces between the components and supporting structures. In traditional density-based topology optimization such as the SIMP model-based method, it is known that the pseudo-density variables used to describe the material layout are always defined with respect to a fixed finite element mesh, i.e. one pseudo-density is related to one element. However, for an integrated layout and topology optimization problem, components' positions will be designed simultaneously with the material layout of the supporting structure. Consequently, the finite element mesh of the design domain has to be updated iteratively to follow the variations of the component positions in a body-fit way.

Suppose an intermediate topology pattern is generated as shown in Figure 5.9. The black and white colors denote the solid and void material properties, respectively. The optimization aims to update the geometry variables and pseudo-densities simultaneously. When the locations and orientations of the components are changed during the iteration, a new finite element mesh has to be generated to make sure the components are still embedded and joined together with the structure. However, after the remeshing, the pseudo-densities cannot find the previous corresponding elements and the structural material layout cannot be further updated.

Accordingly, the method of density points is proposed to solve this confliction. The idea is to relate the pseudo-densities with the points rather than elements. To do this, some fixed points named density points are first defined in the design domain and the pseudo-densities are then attributed to these points. The material properties will be, in turn, spread out from these points to the neighborhood elements.

The method is as follows. First, the density points are defined in the design domain to cover the overall area. As shown in Figure 5.10, four density points are defined to describe four material properties in a square area. Later, the design domain is meshed with elements of proper sizes. By calculating the distances between each density point and the centroids of the corresponding elements, each element will find the nearest density point and receive the information of the material properties. Note that several elements may share the same material property.

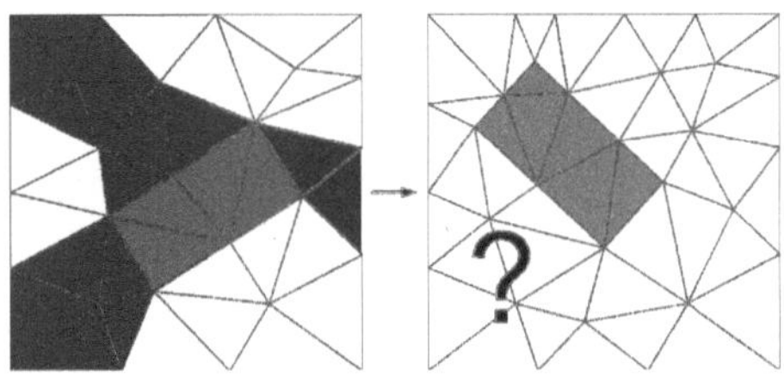


**Figure 5.9.** *Conflict between pseudo-density definition and component position updating*

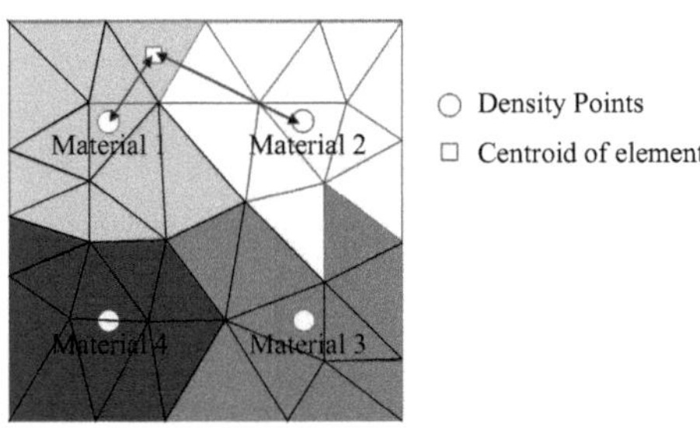


**Figure 5.10.** *Material properties defined by the density points*

With the technique of density points, the finite element mesh can be updated during the iteration of topology optimization. The drawbacks of this kind of element mesh are, however, twofold. On the one hand, a large member of elements will be generated in the finite element model. On the other hand, the whole system has to be remeshed along with the perturbations of the geometry variables and the design iteration. Both of them will cost much computing time. For this reason, the technique of embedded meshing is introduced to embed the components of designable locations and orientations in the design domain with fewer elements and easier mesh updating. As shown in Figure 5.11, the initial mesh of the design domain referred to as the basic mesh or background mesh is first created by fine quadrangular elements. The components are also meshed with refined element size.

When the component is located in the design domain, as shown in Figure 5.12(a), Boolean operations have to be carried out in such a way that some elements of the basic mesh overlapping with the component, as indicated with gray color, will be refined locally to ensure the elements of the component are embedded in the design domain. The modified elements belonging to the design domain will also be restricted in the small square elements of the basic mesh as shown in Figure 5.12(b). Material properties of these elements will still be covered by the proper density points except those belonging to the components. In this way, only a few density points that are located around the component will cover more than one element, which avoids using a large number of the elements to mesh the whole structural system. Furthermore, when the component changes its location and orientation as shown in Figure 5.12(c), the basic mesh is simply restored and only the Boolean operations and modification of the affected elements will be repeated at the new position, rather than remeshing the global system. The final element mesh is shown in Figure 5.12(d).

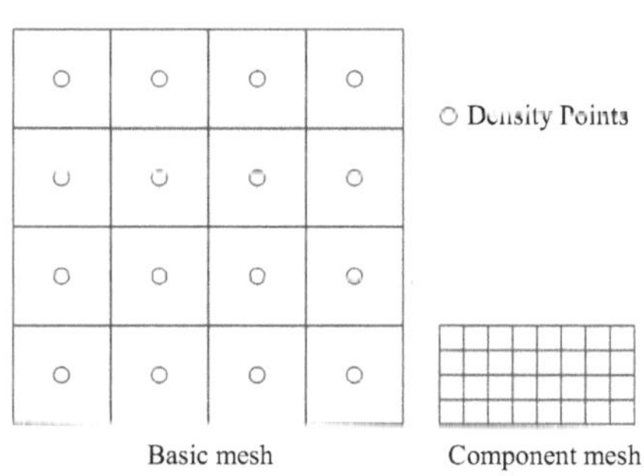


**Figure 5.11.** *Basic mesh with density points and the component mesh*

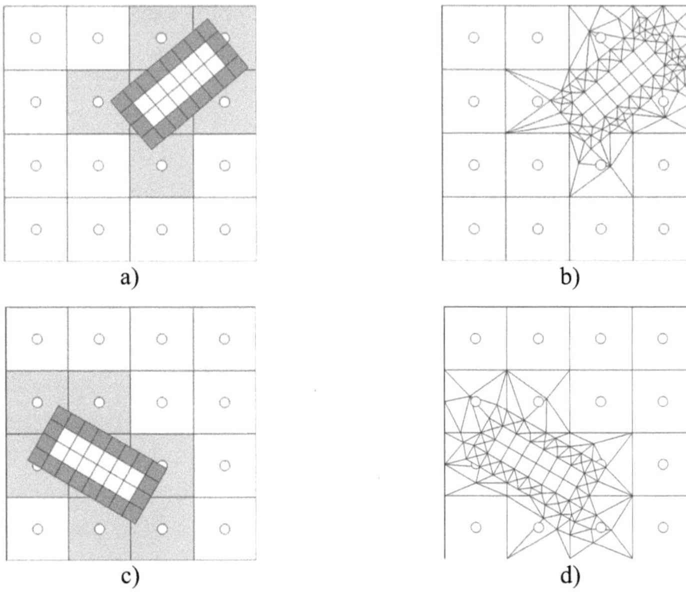


**Figure 5.12.** *Process of embedded meshing*

### 5.3.2. *Superelement and semi-analytical sensitivities*

To enhance the design efficiency, components can generally be modeled as a superelement whose stiffness matrix can be analytically calculated as a function of its orientation. Only the DOFs associated with boundary nodes of the component are retained, while the interior DOFs are condensed. The reduced stiffness matrix is thus expressed as

$$\overline{\mathbf{K}}_{\psi} = \mathbf{K}_{bb} - \mathbf{K}_{bi}\mathbf{K}_{ii}^{-1}\mathbf{K}_{ib} \qquad [5.28]$$

Note that the submatrices correspond to the following partitioned matrix form of the boundary and interior DOFs

$$\mathbf{K}_{\psi} = \begin{bmatrix} \mathbf{K}_{ii} & \mathbf{K}_{ib} \\ \mathbf{K}_{bi} & \mathbf{K}_{bb} \end{bmatrix} \qquad [5.29]$$

Clearly, when the component location is modified by translation in the design procedure, the stiffness matrix of the superelement remains unchanged. In contrast, when the design modification concerns the change of its orientation angle $\theta$, the stiffness matrix can be analytically calculated in a cost-effective way by means of the following transformation:

$$\overline{\mathbf{K}}_{\psi}(\theta) = \mathbf{T}^{\mathrm{T}}(\theta)\overline{\mathbf{K}}_{\psi}^{*}\mathbf{T}(\theta) \qquad [5.30]$$

Meanwhile, the stiffness matrix of the superelement can be calculated only once for all the identical components.

During the sensitivity analysis, the global stiffness is partitioned into three parts contributed from connecting structure, transitional structures and components;

$$\mathbf{K} = \mathbf{K}_{(S)} + \sum_{\psi=1}^{n_c} \mathbf{K}_{(T)\psi} + \sum_{\psi=1}^{n_c} \mathbf{K}_{\psi} \qquad [5.31]$$

$\mathbf{K}_{(S)}$ is the stiffness matrix of the connecting structure, $\mathbf{K}_{(T)\psi}$ is the transition stiffness matrix of the $\psi$th component and $\mathbf{K}_{\psi}$ is the stiffness matrix of the $\psi$th component.

The differentiation of the finite element static equation with respect to $s_j$ of component $j$ gives

$$\left(\frac{\partial \mathbf{K}_{(S)}}{\partial s_j} + \sum_{\psi=1}^{n_c} \frac{\partial \mathbf{K}_{(T)\psi}}{\partial s_j} + \sum_{\psi=1}^{n_c} \frac{\partial \mathbf{K}_{\psi}}{\partial s_j}\right)\mathbf{U} + \left(\mathbf{K}_{(S)} + \sum_{\psi=1}^{n_c} \mathbf{K}_{(T)\psi} + \sum_{\psi=1}^{n_c} \mathbf{K}_{\psi}\right)\frac{\partial \mathbf{U}}{\partial s_j} = 0 \qquad [5.32]$$

While the derivative of the system strain energy with respect to $s_j$ is expressed as

$$\begin{aligned}\frac{\partial C}{\partial s_j} &= \frac{1}{2}\mathbf{f}^T \frac{\partial \mathbf{U}}{\partial s_j} = \frac{1}{2}\mathbf{U}^T\left(\mathbf{K}_{(S)} + \sum_{l=1}^{n_c} \mathbf{K}_{(T)l} + \sum_{l-1}^{n_c} \mathbf{K}_{(C)l}\right)\frac{\partial \mathbf{U}}{\partial s_j} \\ &= -\frac{1}{2}\mathbf{U}^T\left(\frac{\partial \mathbf{K}_{(S)}}{\partial s_j} + \sum_{l=1}^{n_c} \frac{\partial \mathbf{K}_{(T)l}}{\partial s_j} + \sum_{l=1}^{n_c} \frac{\partial \mathbf{K}_{(C)l}}{\partial s_j}\right)\mathbf{U}\end{aligned} \qquad [5.33]$$

$\mathbf{K}_{(S)}$ is independent of $s_j$. The derivative term is then equal to zero. For $\mathbf{K}_{(T)\psi}$, the derivative can be calculated by finite difference so that

$$\sum_{\psi=1}^{n_c} \frac{\partial \mathbf{K}_{(T)\psi}}{\partial s_j} = \sum_{\psi=1}^{n_c} \frac{\mathbf{K}^*_{(T)\psi} - \mathbf{K}_{(T)\psi}}{\Delta s_j} \qquad [5.34]$$

For $\mathbf{K}_\psi$, it is only sensitive to the rotation angle $\theta_j$ of its own, i.e. when $j=\psi$. The derivative of $\mathbf{K}_\psi$ can be simplified as

$$\sum_{l=1}^{n_c} \frac{\partial \mathbf{K}_\psi}{\partial s_j} = \begin{cases} 0 \quad , \quad s_j \neq \theta_j \\ \dfrac{\partial \mathbf{T}_j(\theta_j)^T}{\partial \theta_j} \overline{\mathbf{K}}_j \mathbf{T}_j(\theta_j) + \mathbf{T}_j(\theta_j)^T \overline{\mathbf{K}}_j \dfrac{\partial \mathbf{T}_j(\theta_j)}{\partial \theta_j}, \quad s_j = \theta_j \end{cases} \qquad [5.35]$$

The sensitivity of global strain energy with respect to the pseudo-density variables can be derived analytically, which can be found in many existing literatures and will not be provided here.

Consider a rectangular domain of 0.4 m × 0.5 m containing nine identical rectangular components of 0.05 m × 0.1 m, as shown in Figure 5.13. The rectangular domain is meshed with 40 × 50 four-node quadrilateral membrane elements and each component is modeled as a superelement containing 25 × 50 four-node membrane elements. Two point-wise forces and a moment approximated by a pair of equal and opposite forces are applied on the structure. The right part of the upper edge and the lower part of the left edge are fixed. To avoid possible overlapping during the optimization, each component is approximated with eight circles. Due to the presence of nine components, the volume fraction of the domain is limited to 27.5%. In this way, the final configuration is mainly composed of the components. Here, the elastic modulus of the design domain and the components are $7 \times 10^{10}$ Pa and $2 \times 10^{11}$ Pa, respectively.

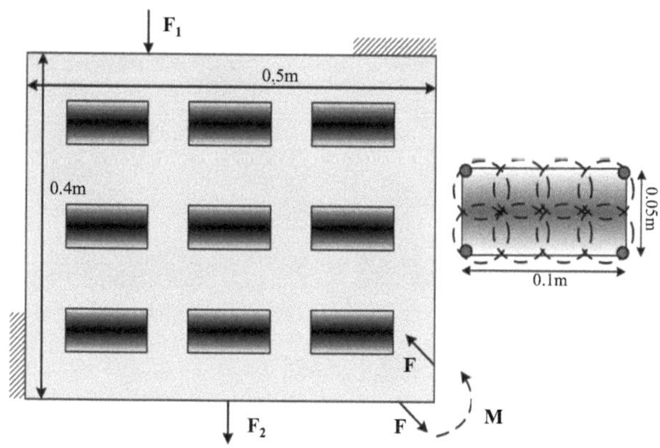


**Figure 5.13.** *Integrated optimization of a nine-component problem*

The evolution of the design configuration is shown in Figure 5.14. Nine components move inside the design domain together with the progressive appearance of the structural framework. The final optimized design is shown in Figure 5.15(a). For the purpose of comparison, Figure 5.15(b) shows the optimized configuration without components, obtained by standard topology optimization with a 50% volume fraction. It is observed that the components' packing design and the structure topology design are integrated simultaneously.

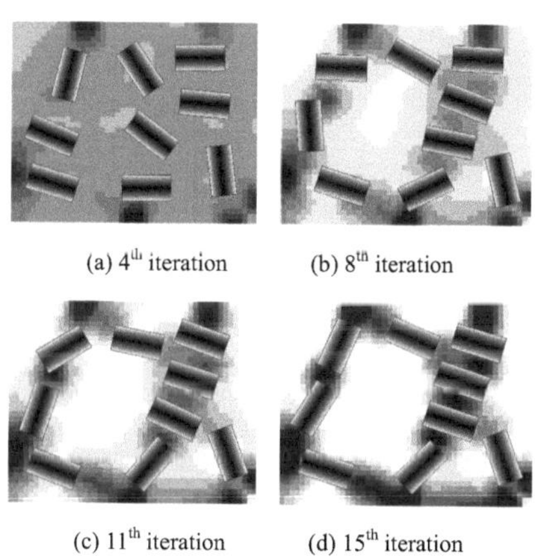


**Figure 5.14.** *Design iterations of a nine-component system*

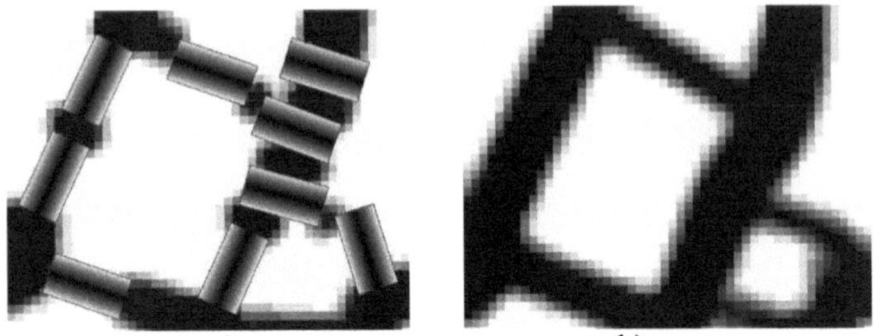


**Figure 5.15.** *Comparison of the integrated optimization and the standard optimization: a) integrated optimization, C = 28.4 J; b) standard optimization, C = 34.6 J*

### 5.3.3. *Decomposition optimization strategies*

Considering the nature of the integrated optimization, decomposition optimization strategies are used by means of the Kuhn–Tucker optimality condition to carry out the packing design and topology design separately, which makes it possible to explore better optimization solutions. In packing optimization, only geometry variables are optimized iteratively, while the pseudo-density design variables are considered as fixed parameters and will be updated in topology optimization. Likewise, in topology optimization, only pseudo-density variables are optimized iteratively with the location design variables fixed. In this context, two decomposition formulations, i.e. parallel design strategy and sequential design strategy, are tested.

In order to have the feature of the considered optimization problem, the formulation can be written in a general form

$$\begin{aligned} &\min: f(\boldsymbol{\eta},\mathbf{S}) \\ &s.t.: \mathbf{h}(\boldsymbol{\eta}) \le \mathbf{0},\ \mathbf{g}(\mathbf{S}) \le \mathbf{0} \end{aligned} \qquad [5.36]$$

This formulation is characterized by the fact that two kinds of design variables are separated at the level of constraints but coupled by the objective function. Based on the Kuhn–Tucker optimality condition, the corresponding Lagrangian function is expressed as

$$L(\boldsymbol{\eta},\mathbf{S},\boldsymbol{\mu},\boldsymbol{\lambda}) = f(\boldsymbol{\eta},\mathbf{S}) + \boldsymbol{\mu}^T\mathbf{h}(\boldsymbol{\eta}) + \boldsymbol{\lambda}^T\mathbf{g}(\mathbf{S}) \quad [5.37]$$

According to the stationary condition, it follows that

$$\begin{aligned} \frac{\partial L}{\partial \boldsymbol{\eta}} &= \frac{\partial f}{\partial \boldsymbol{\eta}} + \boldsymbol{\mu}^T \frac{\partial \mathbf{h}}{\partial \boldsymbol{\eta}} = 0 \\ \frac{\partial L}{\partial \mathbf{S}} &= \frac{\partial f}{\partial \mathbf{S}} + \boldsymbol{\lambda}^T \frac{\partial \mathbf{g}}{\partial \mathbf{S}} = 0 \end{aligned} \quad [5.38]$$

The satisfaction of optimality conditions implies that the original optimization problem is split into two subproblems and each of them contains only one kind of design variable.

$$\begin{aligned} &\min : f_1(\boldsymbol{\eta}) = f(\boldsymbol{\eta},\mathbf{S}_0) \\ &s.t.: \mathbf{h}(\boldsymbol{\eta}) \le 0 \\ \\ &\min : f_2(\mathbf{S}) = f(\boldsymbol{\eta}_0,\mathbf{S}) \\ &s.t.: \mathbf{g}(\mathbf{S}) \le 0 \end{aligned} \quad [5.39]$$

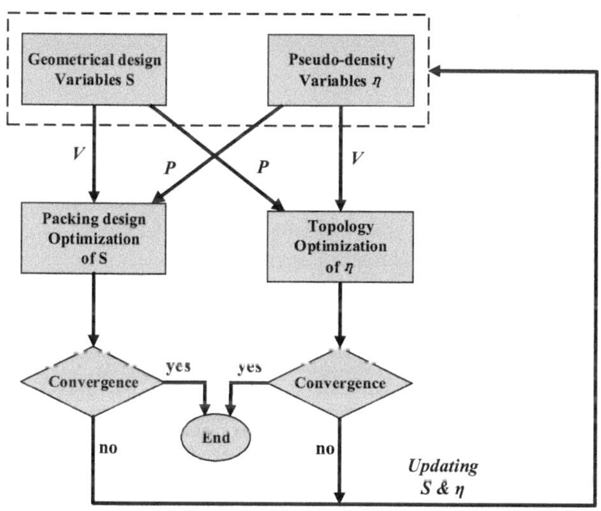


**Figure 5.16.** *Parallel design strategy (V: variables, P: parameters)*

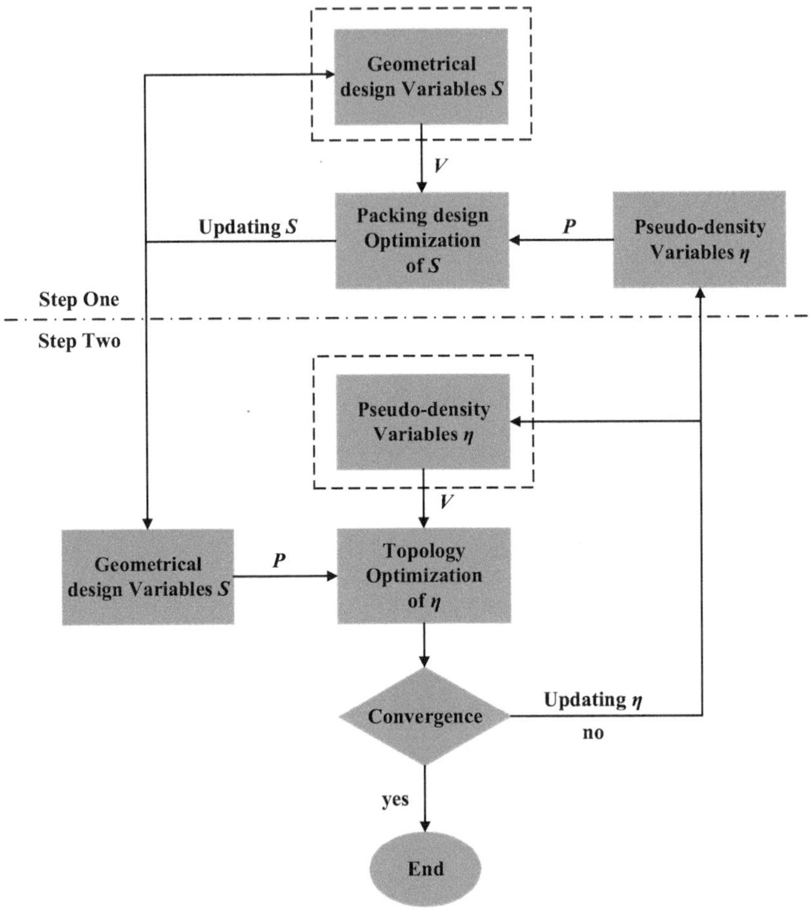


**Figure 5.17.** *Sequential design strategy (V: variables, P: parameters)*

The advantage of this decomposition is twofold. On the one hand, the problem scale can be reduced to some extent and each subproblem can be solved independently at the current step. On the other hand, proper approximation concepts and optimization algorithms can be applied independently for pure topology optimization and packing design, because the nature of each set of design variables is different.

This is beneficial in improving the computing efficiency and iteration convergence. Here, two design strategies, i.e. parallel strategy and sequential strategy, are tested in order to save computing time and achieve high-quality convergence of optimization. Parallel strategy refers to the parallel running of two subproblems independently. The flowchart is illustrated in Figure 5.16.

In contrast, the flowchart of sequential design strategy is illustrated in Figure 5.17. The main difference from the parallel strategy is that two kinds of design variables are updated sequentially.

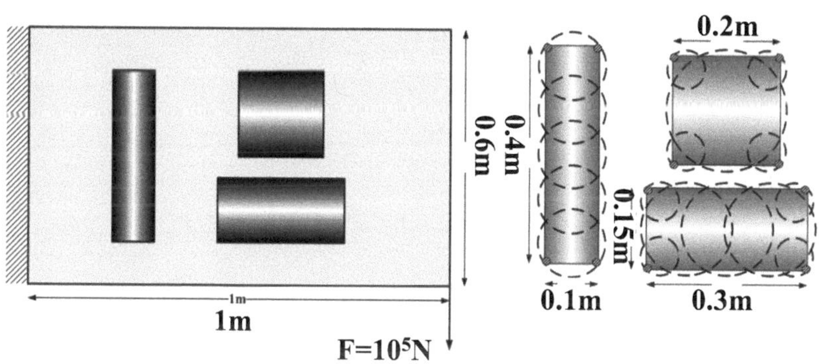


**Figure 5.18.** *Cantilever beam with three components*

A cantilever beam with three components is presented here to investigate different design strategies, i.e. simultaneous design, parallel design and sequential design. As shown in Figure 5.18, the sizes of the three components are different. Material properties are the same as in the previous examples.

As shown in Figure 5.19, the optimized designs achieved by three design strategies are different in the component layout, while the final values of structural strain energy are quite close to each other. This confirms the existence of multiple minima for this kind of NP-hard problem.

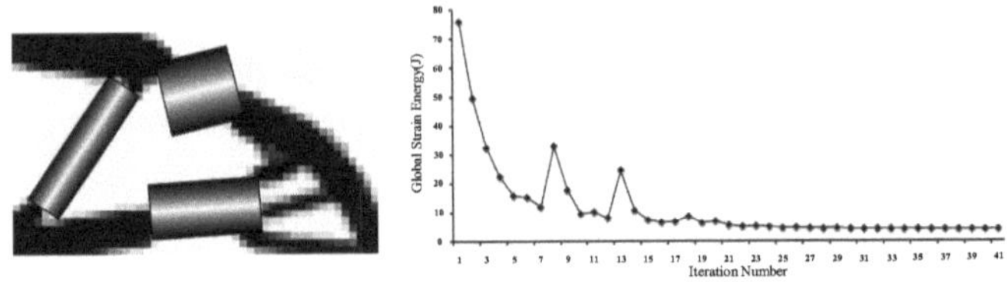


(a) Simultaneous design strategy ($C$=3.84 J)

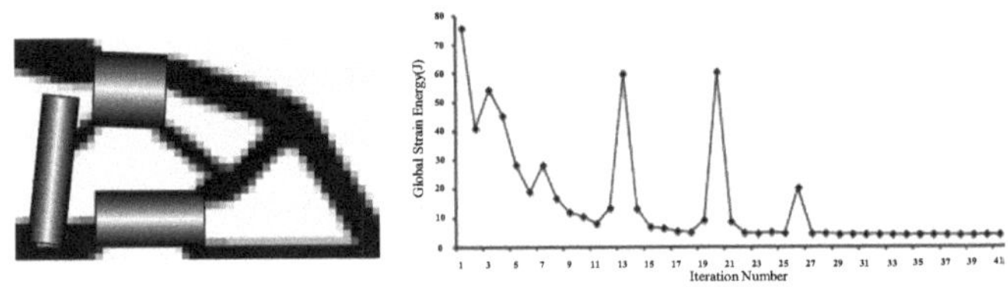


(b) Parallel design strategy ($C$=3.87 J)

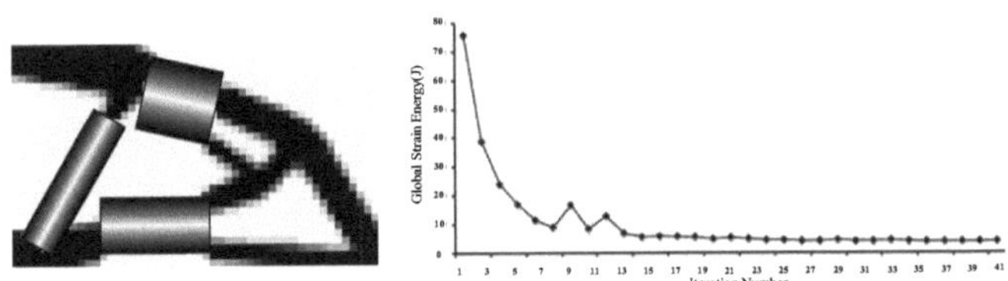


(c) Sequential design strategy ($C$=3.83 J)

**Figure 5.19.** *Optimized designs obtained with three design strategies*

Besides, it is shown in Figure 5.19(a) that about 40 iterations are required by the simultaneous strategy and oscillations take place at iteration 8 and iteration 13. This is because few materials are available to connect the shortest component as shown in Figure 5.20(a). The values of local pseudo-density variables covered by the component reach the lower bound instantly due to its movement and they are quickly recovered after several iterations. After the 13th iteration, only slight changes occur for the component positions so that the iteration history becomes stable.

For the parallel design strategy, nearly the same iteration number is required. However, oscillations become more serious because the coupling between two sets of design variables is greatly weakened. Instead, one such coupling is relatively strengthened by the sequential design strategy so that the corresponding iteration history shown in Figure 5.19(c) converges smoothly.

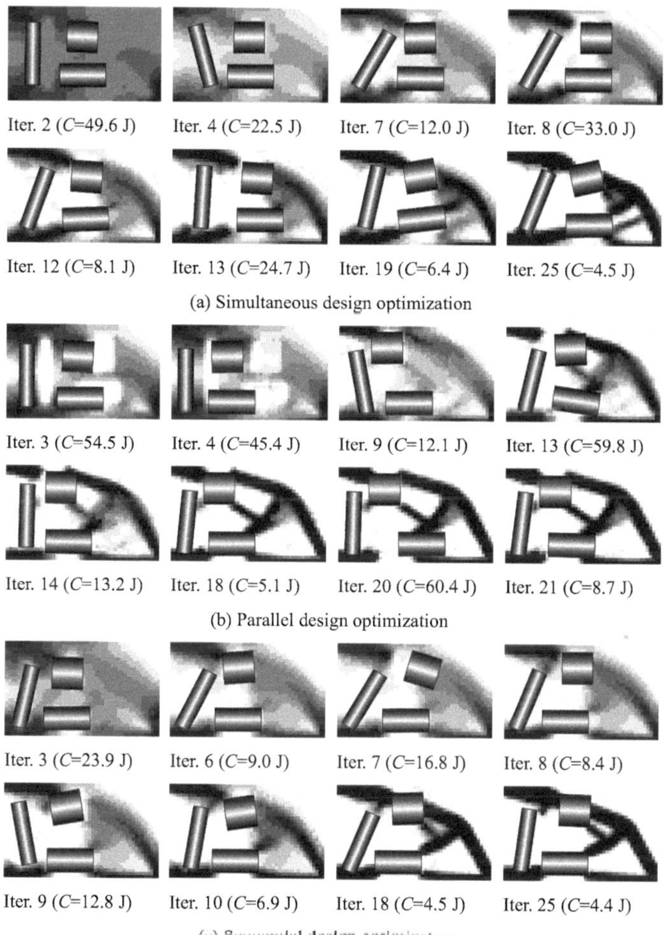


**Figure 5.20.** *Optimization iterations with three design strategies*

## 5.4. MPC-based component-structure connections

In many practical industrial cases, components and supporting structures are mostly assembled together by rivets or bolts with the connecting positions designated in advance as shown in Figure 5.21.

The components are actually floating on the surface of the topological design domain and direct nodal connections are not applicable. Here, we propose to use multipoint constraints (MPCs) to define rivet or bolt connections. The displacement consistence is strictly maintained by satisfying the MPC equations. When the components move, only the MPC connections are rebuilt at new positions. In this way, the advantages of remesh-free, analyticity as well as the precise material description, are simultaneously maintained.

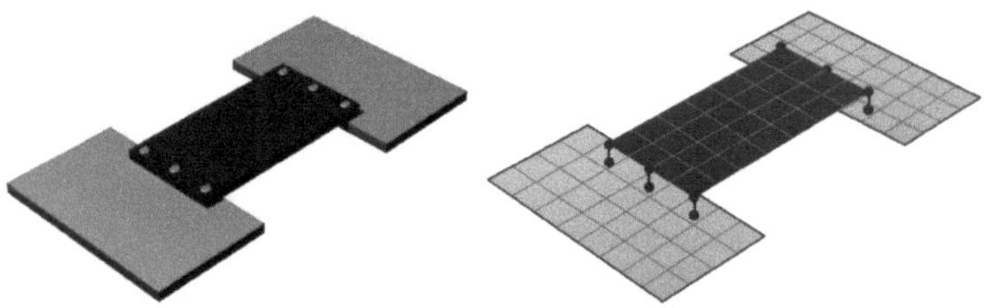

**Figure 5.21.** *Connections between components and supporting structures*

To use MPC as the connections, the topological design domain and the components are discretized, respectively, into finite elements as illustrated in Figure 5.22. Assume $\mathbf{M}_1$ is one of the connecting nodes on the component, which is projected to the point $\mathbf{M}_1^*$ inside the structural element $e_1$. Then, we enable the following MPC equation.

$$\begin{aligned} &\mathbf{U}_{\mathbf{M}1} = \mathbf{U}_{\mathbf{M}1}^* = \mathbf{N}_{e1}\left(\mathbf{M}_1^*\right)\cdot \mathbf{U}_{e1} \\ &\mathbf{U}_{\mathbf{M}1} - \mathbf{N}_{e1}\left(\mathbf{M}_1^*\right)\cdot \mathbf{U}_{e1} = \mathbf{0} \end{aligned} \qquad [5.40]$$

where $\mathbf{U}_{\mathbf{M}1}$ and $\mathbf{U}^*_{\mathbf{M}1}$ denote the displacement vectors of node $\mathbf{M}_1$ and point $\mathbf{M}_1^*$.$\mathbf{U}_{e1}$ and $\mathbf{N}_{e1}(\mathbf{M}_1^*)$ are the displacement vectors of the element $e_1$ and the shape function coefficient matrix at the point $\mathbf{M}_1^*$.

Note that the MPC equation is a linear combination of the nodal displacements. Multiple MPC equations as well as the boundary conditions can be organized as the following linear equations.

$$\mathbf{HU} = \mathbf{0} \qquad [5.41]$$

where **H** is the coefficient matrix determined by the shape functions of the connected elements, connection positions and boundary conditions. **U** is the global displacement vector.

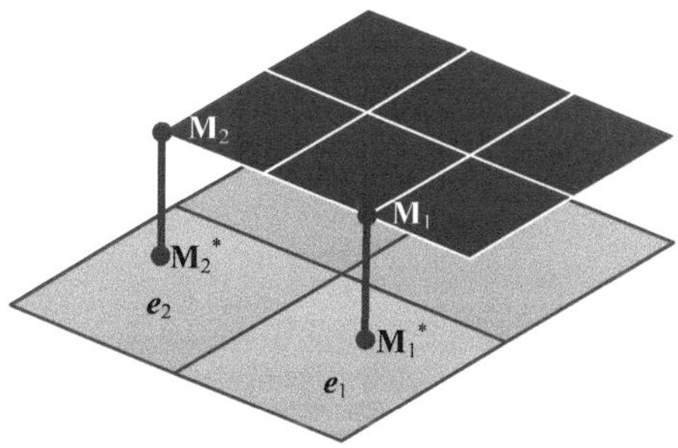


**Figure 5.22.** *Definition of MPC connections*

Considering the above displacement constraints, the revised form of the overall potential energy of the global system can be expressed as:

$$\Pi(\mathbf{U},\boldsymbol{\lambda}) = \frac{1}{2}\mathbf{U}^{\mathrm{T}}\mathbf{K}\mathbf{U} - \mathbf{F}^{\mathrm{T}}\mathbf{U} + \boldsymbol{\lambda}^{\mathrm{T}}\mathbf{H}\mathbf{U} \quad [5.42]$$

where **K** and **F** are, respectively, the global stiffness matrix and the global nodal load vector. $\boldsymbol{\lambda}$ is the Lagrange multiplier vector. The stiffness matrix can be expressed as

$$\mathbf{K} = \begin{bmatrix} \mathbf{K}_s & \mathbf{0} & \mathbf{0} & \mathbf{0} & \mathbf{0} \\ \mathbf{0} & \mathbf{K}_1 & \mathbf{0} & \mathbf{0} & \mathbf{0} \\ \mathbf{0} & \mathbf{0} & \mathbf{K}_2 & \mathbf{0} & \mathbf{0} \\ \mathbf{0} & \mathbf{0} & \mathbf{0} & \ddots & \mathbf{0} \\ \mathbf{0} & \mathbf{0} & \mathbf{0} & \mathbf{0} & \mathbf{K}_{n_c} \end{bmatrix} \quad [5.43]$$

$\mathbf{K}_s$ is the stiffness matrix of the supporting structure. $\mathbf{K}_1$, $\mathbf{K}_2$ and $\mathbf{K}_{nc}$, respectively, stand for the stiffness matrices of the first, second and $n_c$th components. Similar definitions are used for the global load vector and displacement vector. Then, we apply the stationary conditions and obtain:

$$\begin{cases} \mathbf{KU} + \mathbf{H}^{\mathrm{T}}\boldsymbol{\lambda} = \mathbf{F} \\ \mathbf{HU} = \mathbf{0} \end{cases} \qquad [5.44]$$

By solving the above equations, the displacement vector $\mathbf{U}$ and Lagrange multiplier vector $\boldsymbol{\lambda}$ can be finally obtained. Similar explanations of MPC equations can be found in some existing works such as Ainsworth [AIN 01] and Yoon *et al.* [YOO 04]. Typically, to move the components during the optimization iteration, we only have to relocate the finite element models of the components at the new positions and rebuild the MPC equations. Since there are no direct nodal connections between the components and the supporting structures, the element remeshing is avoided in this procedure.

Considering the sensitivity analysis, the differentiation of the static equation with respect to the pseudo-density variable $\eta_i$ can be written as:

$$\frac{\partial \mathbf{K}}{\partial \eta_i}\mathbf{U} + \mathbf{K}\frac{\partial \mathbf{U}}{\partial \eta_i} = \frac{\partial(\mathbf{F} - \mathbf{H}^{\mathrm{T}}\boldsymbol{\lambda})}{\partial \eta_i} \qquad [5.45]$$

Assuming $\mathbf{F}=\mathbf{F}_{\mathrm{m}}+\mathbf{F}_{\mathrm{g}}$, where $\mathbf{F}_{\mathrm{m}}$ and $\mathbf{F}_{\mathrm{g}}$ stand for design independent external loads and design dependent inertial loads, respectively, we have:

$$\frac{\partial(\mathbf{F} - \mathbf{H}^{\mathrm{T}}\boldsymbol{\lambda})}{\partial \eta_i} = \frac{\partial \mathbf{F}_{\mathrm{g}}}{\partial \eta_i} - \frac{\partial(\mathbf{H}^{\mathrm{T}}\boldsymbol{\lambda})}{\partial \eta_i} \qquad [5.46]$$

Then, we have the derivative of the overall strain energy:

$$\begin{aligned} \frac{\partial C}{\partial \eta_i} &= \mathbf{U}^{\mathrm{T}}\mathbf{K}\frac{\partial \mathbf{U}}{\partial \eta_i} + \frac{1}{2}\mathbf{U}^{\mathrm{T}}\frac{\partial \mathbf{K}}{\partial \eta_i}\mathbf{U} \\ &= \mathbf{U}^{\mathrm{T}}\frac{\partial \mathbf{F}_{\mathrm{g}}}{\partial \eta_i} - \mathbf{U}^{\mathrm{T}}\mathbf{H}^{\mathrm{T}}\frac{\partial \boldsymbol{\lambda}}{\partial \eta_i} - \frac{1}{2}\mathbf{U}^{\mathrm{T}}\frac{\partial \mathbf{K}}{\partial \eta_i}\mathbf{U} \\ &= \mathbf{U}^{\mathrm{T}}\frac{\partial \mathbf{F}_{\mathrm{g}}}{\partial \eta_i} - \frac{1}{2}\mathbf{U}^{\mathrm{T}}\frac{\partial \mathbf{K}}{\partial \eta_i}\mathbf{U} \end{aligned} \qquad [5.47]$$

where the derivatives of the inertial load vector $\mathbf{F}_g$ and the stiffness matrix $\mathbf{K}$ can be easily obtained according to the material interpolation model for the element mass and stiffness.

The derivative of the strain energy with respect to the geometry design variable $s_j$ is similarly written as:

$$\frac{\partial C}{\partial s_j} = -\mathbf{U}^{\mathrm{T}} \frac{\partial \mathbf{H}^{\mathrm{T}}}{\partial s_j} \boldsymbol{\lambda} - \frac{1}{2} \mathbf{U}^{\mathrm{T}} \frac{\partial \mathbf{K}}{\partial s_j} \mathbf{U} \qquad [5.48]$$

Suppose $s_j$ is a translational variable of the $j$th component, both the stiffness matrix of the supporting structure and the components will remain unchanged after a translational moving. So, we yield:

$$\frac{\partial C}{\partial s_j} = -\mathbf{U}^{\mathrm{T}} \frac{\partial \mathbf{H}^{\mathrm{T}}}{\partial s_j} \boldsymbol{\lambda} \qquad [5.49]$$

If $s_j$ is a rotational variable, the derivative of the stiffness matrix can be done in a similar way to equations [5.30]–[5.35].

Figure 5.23 illustrates an example. A hexagonal plate with a thickness of 0.036 m is discretized into three layers of total 18,000 solid hexahedron elements. External loads are applied at three different positions. Each position has four nodes and a 500 N downward force has been applied on each single node. The design domain is described with six large circles by FCM. The elastic modulus and the Poisson's ratio are $7.0 \times 10^{10}$ Pa and 0.3.

The first two identical six-foot components are approximately described by four circles. Their elastic modulus and the Poisson's ratio are $1.1 \times 10^{11}$ Pa and 0.3. The elastic modulus and the Poisson's ratio of the second group of two four-foot components described by three circles are $2.0 \times 10^{11}$ Pa and 0.3. Detailed configurations of the components are illustrated in Figure 5.24 and all four components are connected to the supporting structure surface through the nodes in their foot areas.

There are in total 109 non-overlapping constraints aggregated by the improved adaptive method based on KS function in this problem. The material volume fraction is constrained to 0.4.

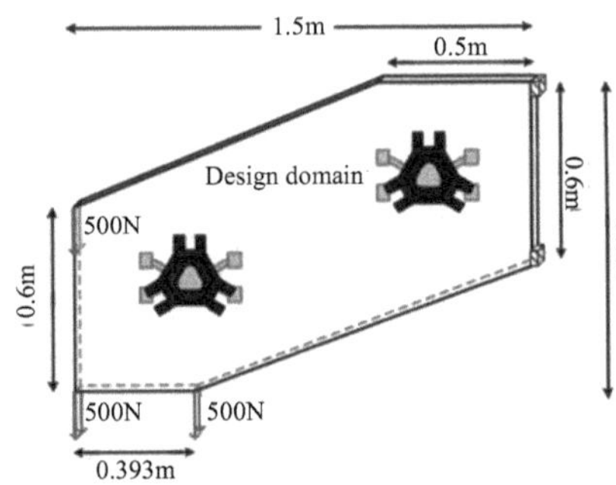


**Figure 5.23.** *Optimization problem with four components*

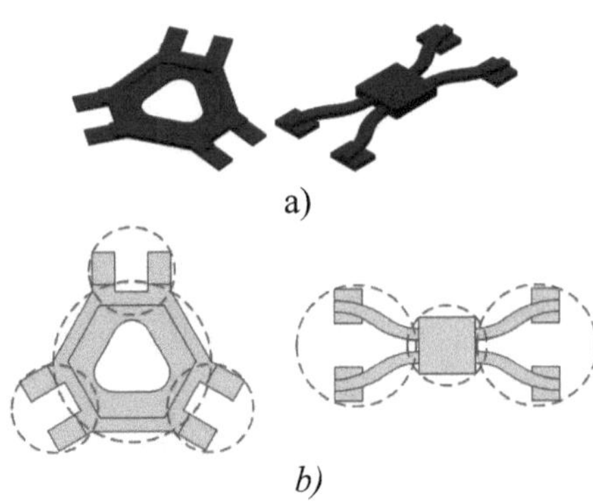


**Figure 5.24.** *Detailed illustrations of the components: a) the 3D models of two sets of components; b) circle approximation by FCM*

As shown in Figures 5.25(a–g), clear structural topologies are obtained and all four components find their proper locations to reinforce the structure locally. The optimization converges after 70 iterations with the global strain energy decreasing smoothly from an initial 11.267 J to a final value of 0.877 J as plotted in Figure 5.26. In the optimized design, it is found that the two four-foot components are relatively ineffective in carrying loads, which ultimately behave as the adhering and reinforcing components. In contrast, the six-foot components are much stiffer and they are thus located on the key load carrying path.

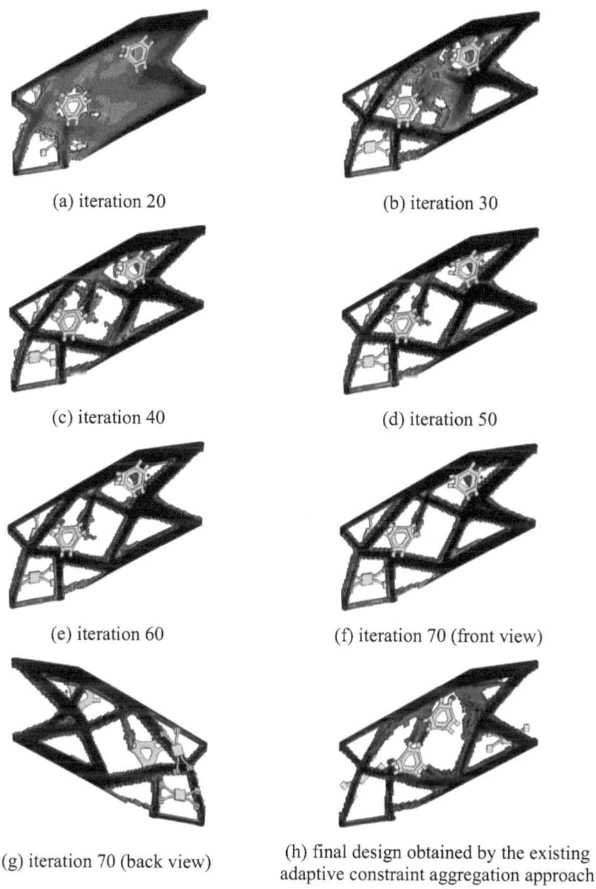

(a) iteration 20
(b) iteration 30
(c) iteration 40
(d) iteration 50
(e) iteration 60
(f) iteration 70 (front view)
(g) iteration 70 (back view)
(h) final design obtained by the existing adaptive constraint aggregation approach

**Figure 5.25.** *Design iterations of structural topology and components' layout*

For the purpose of comparison, this problem is solved again with the existing adaptive constraint aggregation approach [MAR 05, POO 07] based on KS function. The final structural topology shows that the optimization does not converge after 70 iterations, as shown in Figure 5.25(h). Its final global strain energy is also higher than that obtained by the improved approach in Figure 5.26. The final design is also found to be infeasible due to overlapping between one component and the design domain boundary. The iteration histories of aggregated constraints using the improved adaptive approach and the existing adaptive approach are shown in Figure 5.27.

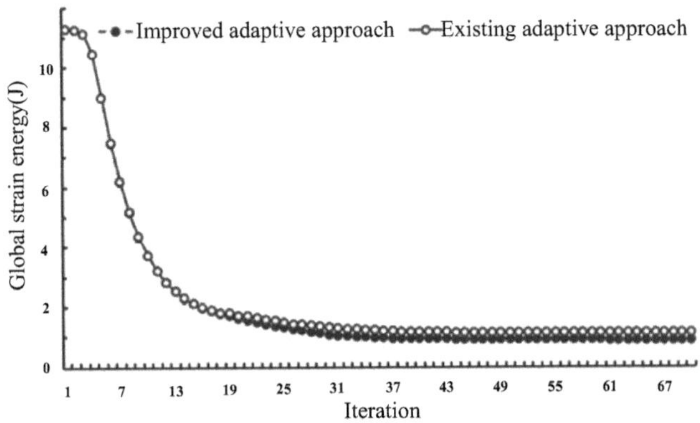


**Figure 5.26.** *Iteration history of the global strain energy*

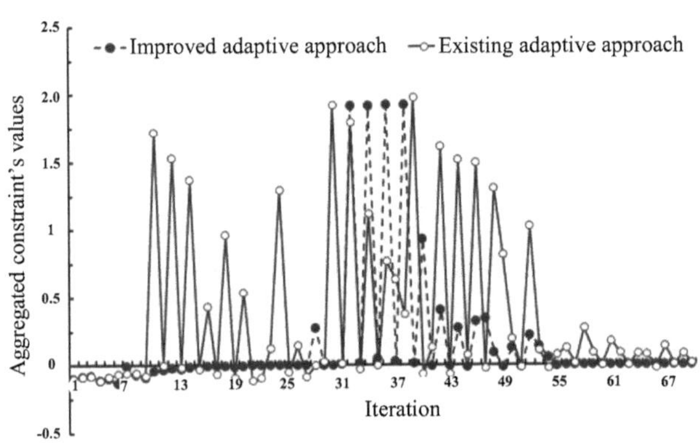


**Figure 5.27.** *Iteration history of aggregated constraint*

The evolutions of aggregation parameters and their derivatives in both adaptive approaches are also shown in Figures 5.28 and 5.29. It is found that the aggregation parameters and the derivatives of aggregated constraints obtained by the improved adaptive approach show a convergent and stable history, while those calculated by the existing approach stay divergent during the whole optimization.

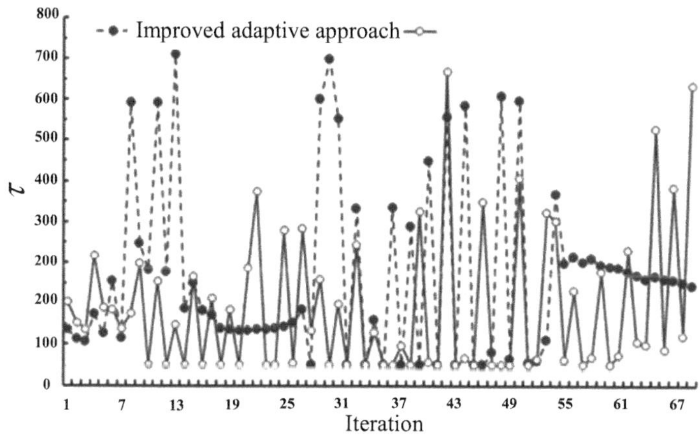


**Figure 5.28.** *Iteration history of aggregation parameters obtained by both approaches*

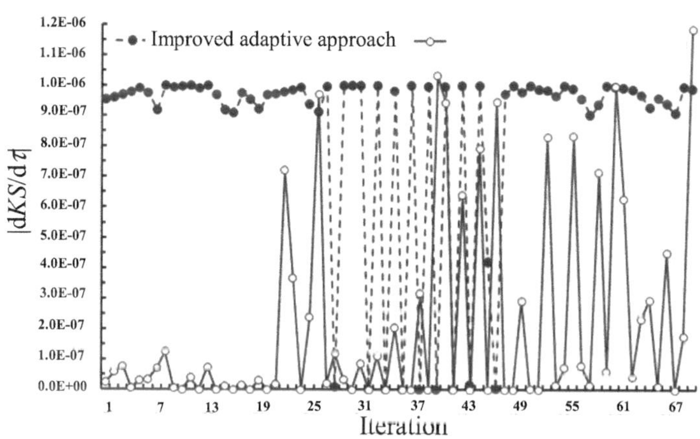


**Figure 5.29.** *Comparison of |dKS/dτ| calculated by both approaches*

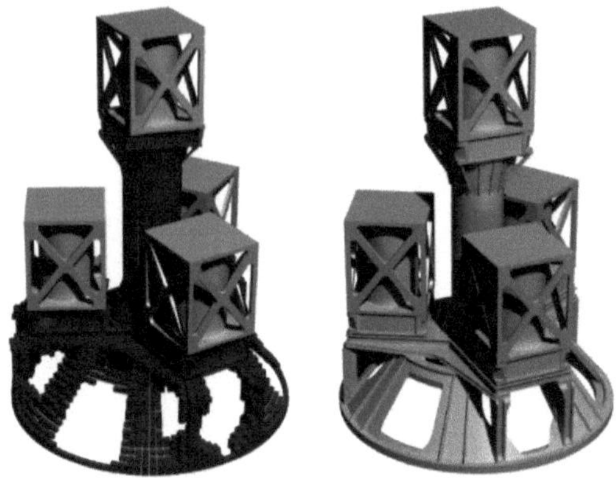

**Figure 5.30.** *Multicomponent layout design of a four-satellite supporting system*

Integrated optimization was recently applied to design a launching vehicle structure which supports four satellites as shown in Figure 5.30. One satellite was located on the top and three satellites were evenly located on a lower plate. The orientation of the top satellite and distances of the lower satellites to the central axis are assigned as the geometry variables. The supporting structure below the plate was assigned as the topological design domain. The objective is to maximize the designated natural frequencies. A prescribed weight limit together with several other design constraints on the structural symmetry, such as inner space for other devices, etc., has been considered in the design. Figure 5.30 shows the topology optimization result and the final engineering design. Compared with the original design, the optimization has increased the fundamental natural frequency by 17% and reduced the weight by 7%.

## 5.5. Integrated optimization based on implicit model

### 5.5.1. *Implicit representation of component geometry*

In topology optimization, the density-based method has been acknowledged as a promising method with countless extensions and industrial applications and is commonly used with different material interpolation schemes. The proposed integrated optimization using density points, embedded meshing and MPC has proved effective in obtaining reasonable optimization design. Here, new schemes based on the implicit model will be presented including improvements in components modeling and sensitivity analysis.

The FE discretization of an integrated system with a circular component and a square host structure illustrated in Figure 5.31(a) can be considered as a local Lagrangian mesh as shown in Figure 5.31(b). The adjacent elements attaching to the moving boundary of the component need to be locally refined by adding transition elements. Although the Lagrangian mesh provides the accuracies of geometry and material distribution, in some case it will fail to prevent poor element qualities and difficulties in mesh generation and design iterations.

Here, we employ a fixed Eulerian mesh, as shown in Figure 5.31(c). As the material interfaces do not coincide with the real geometrical interface, the discontinuity of material properties over adjacent elements attaching to the boundary is smoothed approximately [QIA 04, CHE 07]. Such a regular mesh type automatically favors standard topology optimization related to a fixed FE mesh. Obviously, a refined mesh shown in Figure 5.31(d) will increase the computing accuracy at the cost of computing time.

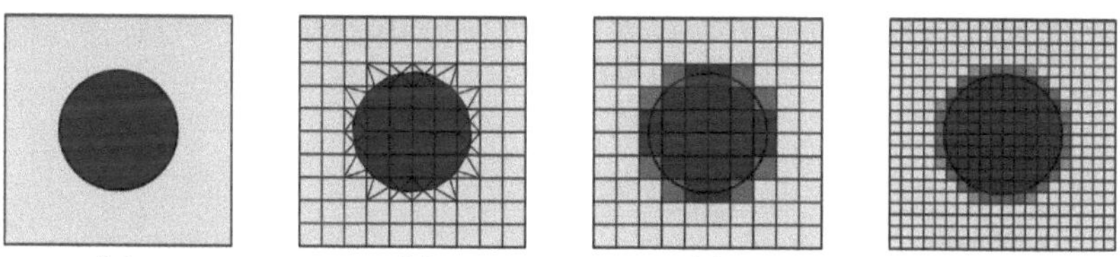

(a) (b) (c) (d)

**Figure 5.31.** *Different meshing strategies: a) initial material layout, b) local Lagrangian mesh, c) Eulerian mesh and d) refined Eulerian mesh*

By means of the *level set* method and R-function theory, the free-form shape of a component can be represented by an implicit function. This is an alternative method to explicit parameterization such as Bezier, Nurbs curves and surfaces for B-reps, and can be well adapted to the Eulerian mesh. As shown in Figure 5.32(a), a 3D *level set* function $\Phi(\mathbf{x})$ is constructed according to the center coordinates and radius of a circular component in Figure 5.32(b). Suppose $\mathbf{x}$ denotes the coordinates of a point in the design domain $\boldsymbol{D}$. $\Phi(\mathbf{x})$ then corresponds to

$$\begin{cases} \mathbf{x} \text{ inside } \Omega, & \text{if } \Phi(\mathbf{x}) > 0 \\ \mathbf{x} \text{ on } \Gamma, & \text{if } \Phi(\mathbf{x}) = 0 \\ \mathbf{x} \text{ outside } \Omega, & \text{if } \Phi(\mathbf{x}) < 0 \end{cases} \qquad [5.50]$$

where $\Phi(\mathbf{x})>0$, $\Phi(\mathbf{x})<0$ and $\Phi(\mathbf{x})=0$ denotes the solid, void and boundary, respectively.

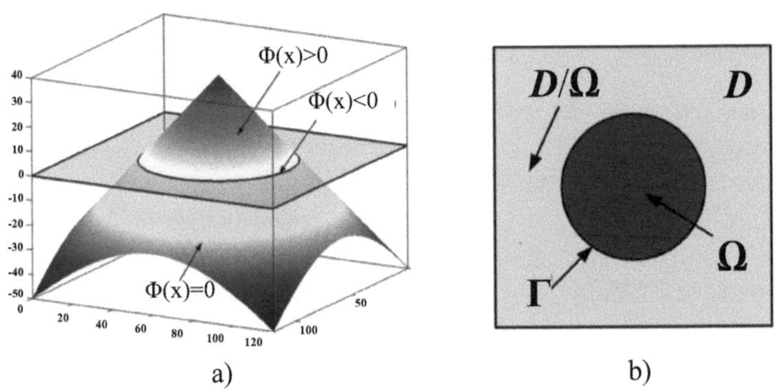


**Figure 5.32.** *Level set representation: a) 3D-level set model; b) 2D design domain*

Likewise, we can resort to R-functions for the implicit representations of complicated boundary shapes. The mathematical formulation is based on the following Boolean operations on constitutive functions.

$$f_{1-2} = \begin{cases} f_1 \cap_0 f_2 = f_1 + f_2 - \sqrt{f_1^2 + f_2^2} \\ f_1 \cup_0 f_2 = f_1 + f_2 + \sqrt{f_1^2 + f_2^2} \end{cases} \qquad [5.51]$$

For example, a rectangular domain $\Omega$ with dimensions of $a \times b$ can be represented by the R-function as

$$\begin{cases} f_{1-2} = f_1 + f_2 - \sqrt{f_1^2 + f_2^2} \\ f_1 = a^2 - x^2 \geq 0 \\ f_2 = b^2 - y^2 \geq 0 \end{cases} \quad [5.52]$$

Such interesting features provide great flexibility in shape representation of components. The component geometry can be decomposed into several basic elements, such as rectangles, ellipses, etc. Through R-function-based Boolean operations, such a complex geometry could be implicitly represented. The following equation gives a possible implicit representation function formation for ellipse:

$$f = 1 - \sqrt{\left(\frac{x - x_c}{a}\right)^2 + \left(\frac{y - y_c}{b}\right)^2} \quad [5.53]$$

where $(x_c, y_c)$ gives the center coordinate of the ellipse and $a$ and $b$ are the radii in two axes.

Theoretically, with the Boolean operations we can implicitly represent shapes of any complexity in required precision with an unlimited number of basic elements. Although the constructed implicit function could be too complicated to be expressed in an exact formulation after several levels of Boolean operations, its exact formulation is not a necessity in the optimization design. Suppose $\Phi_\psi$ is an implicit function representing the higher dimensional shape representation function of the $\psi$th component. $\mathbf{S}_\psi$ represents the set of involved shape parameters, e.g. length, width, radius, etc. Such an implicit representation makes it possible to construct higher dimensional shape representation functions analytically.

For the structural elements in the design domain, we take the SIMP model penalizing the material properties, for example:

$$\begin{cases} E_i = (\eta_i)^p E_0 \\ v_i = \eta_i v_0 \end{cases} \quad [5.54]$$

where $\eta_i$ is the pseudo-density variable of the element $i$, and $p$ is the penalty factor (typically $p = 3$). $E_i$ is the element Young's modulus and $E_0$ is the Young's modulus of solid material.

For the components, suppose the $\psi$th component has a homogenous Young's modulus $E_\psi$. $\mathbf{x}_i$ denotes the centroid coordinate of element $i$ in the Cartesian coordinate system. As a result, a standard material interpolation model is formulated.

$$E_i = (\eta_i)^p \left( E_0 + \sum_{\psi=1}^{n_c} A(\Phi_\psi(\mathbf{x}_i, \mathbf{S}_\psi))(E_\psi - E_0) \right) \quad [5.55]$$

$A(\Phi_\psi(\mathbf{x}_i,\mathbf{S}_\psi))$ acts as a compressing function to transform values of $\Phi_\psi(\mathbf{x}_i,\mathbf{S}_\psi)$ into the range of 0–1. One of its expressions can be written as:

$$\mathrm{A}(t) = \frac{\arctan(t/\Delta)}{\pi} + \frac{1}{2} \quad [5.56]$$

its derivative w.r.t. $t$ is then analytically expressed as:

$$\frac{\mathrm{dA}(t)}{\mathrm{d}t} = \frac{\Delta}{\pi(\Delta^2 + t^2)} \quad [5.57]$$

Figure 5.33 shows A($t$) and its derivative at different values of $\Delta$. Obviously, $\Delta$ of a small value might lead to instabilities in sensitivity analysis, while a large $\Delta$ may result in imprecise modeling. In fact, $\Delta$ controls the local approximation smoothness of material discontinuity over the boundary region. An alternative choice of the compression function is the modified Heaviside function, which is a piecewise-function but has a similar effect to the above function.

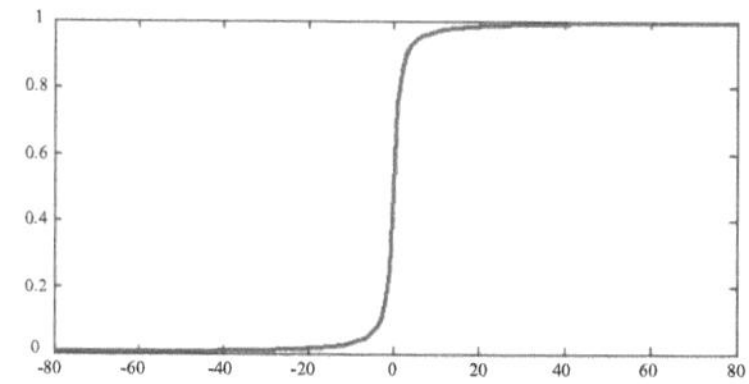


(a) $A(t);\Delta=1$

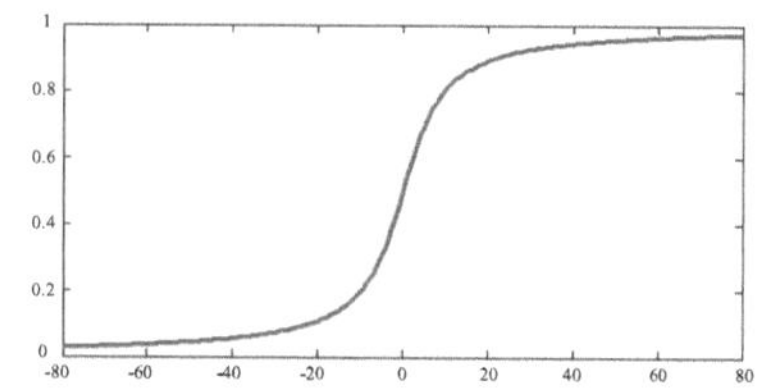


(b) $A(t);\Delta=7$

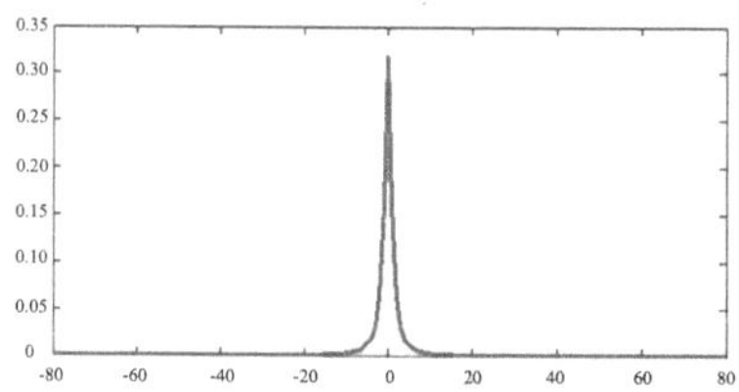


(c) $dA(t)/dt\,;\Delta=1$

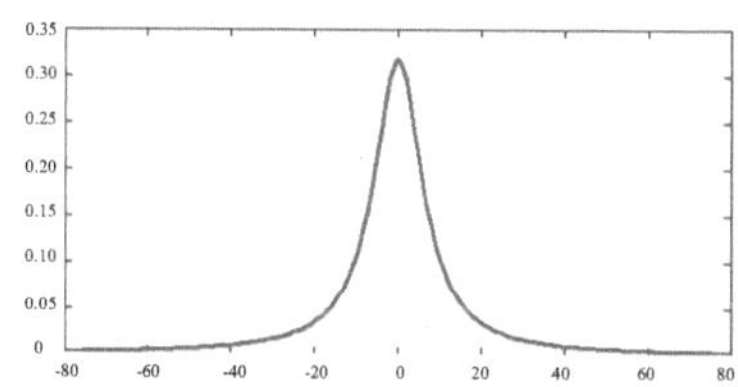


(d) $dA(t)/dt\,;\Delta=7$

**Figure 5.33.** *Curves of A(t) and its derivative of different Δ value with respect to t*

To have a clear idea, the following equations give the *level set* functions representing a circle and a rectangle illustrated in Figures 5.34 and 5.35. Using $\Delta = 4$, the compression functions $A(\Phi)$ are correspondingly shown.

Circle: $\Phi = 24 - \sqrt{x^2 + y^2}$

$$\text{Rectangle: } \Phi = f_1 + f_2 - \sqrt{f_1^2 + f_2^2}$$
$$\text{with } \begin{cases} f_1 = 24 - \sqrt{(x \times \cos 18^\circ - y \times \sin 18^\circ)^2} \\ f_2 = 40 - \sqrt{(x \times \sin 18^\circ + y \times \cos 18^\circ)^2} \end{cases} \qquad [5.58]$$

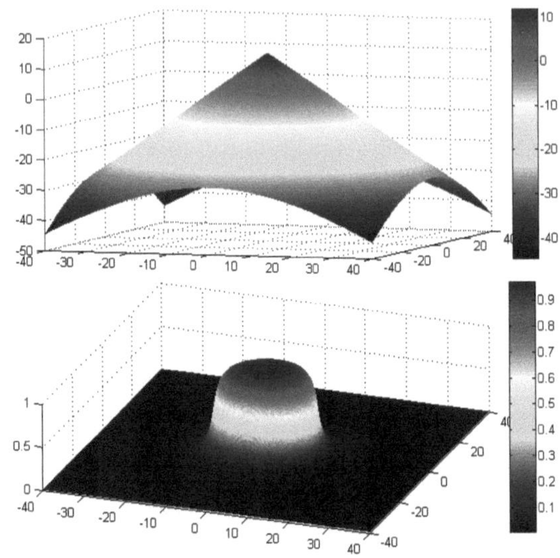

**Figure 5.34.** *Compression of a circle representation function. For a color version of this figure, see www.iste.co.uk/zhang/topology.zip*

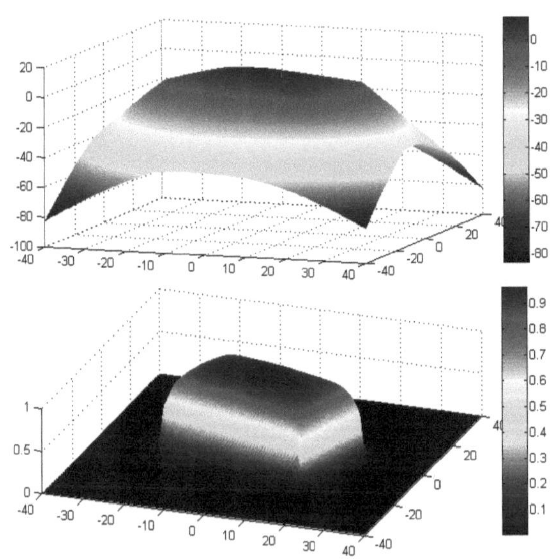

**Figure 5.35.** *Compression of a rectangle representation function. For a color version of this figure, see www.iste.co.uk/zhang/topology.zip*

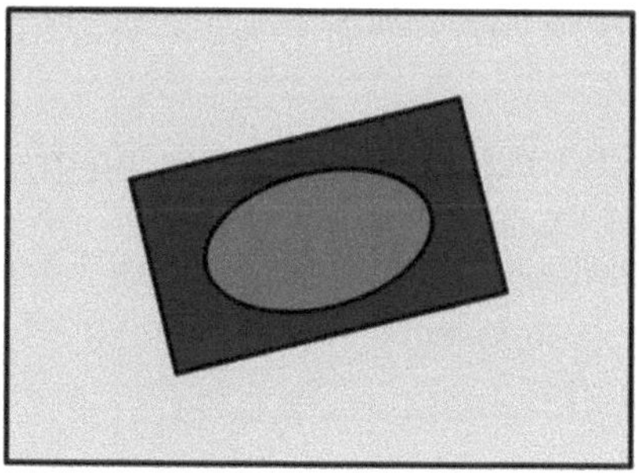


**Figure 5.36.** *Illustration of a component with two material phases*

It is easy to decide the material properties of the elements inside and outside the $\psi$th component, respectively. Notice that each component is assumed to be homogeneous with elasticity $E_\psi$. In fact, equation [5.55] can be generalized to variant material models when multiphase materials are involved in a component. For instance, consider a two-phase component consisting of materials with elastic moduli $E_{\psi 1}$ and $E_{\psi 2}$ as shown in Figure 5.36. $\Phi_{\psi 1}$ and $\Phi_{\psi 2}$ denote the geometry of different material phases in the $\psi$th component. The material model can be defined as:

$$E_i = (\eta_i)^p \left(E_0 + \mathrm{A}(\Phi_{\psi 1}(\mathbf{x}_i, \mathbf{S}_\psi))\left(E_{\psi 1} - E_0\right) + \mathrm{A}(\Phi_{\psi 2}(\mathbf{x}_i, \mathbf{S}_\psi))\left(E_{\psi 2} - E_{\psi 1}\right)\right) \quad [5.59]$$

Clearly, whatever the complexity of the component shape, the number of material phases, and the number of involved components, the material model can be constructed in a generalized manner:

$$E_i = (\eta_i)^p \left( E_0 + \sum_{\psi=1}^{n_c} \sum_{\varsigma=1}^{n_\psi} \mathrm{A}(\Phi_{\psi\varsigma}(\mathrm{x}_i, \mathbf{S}_\psi)) f_\psi \left(E_0, E_{\psi 1}, \ldots, E_{\psi n_\psi}\right) \right) \quad [5.60]$$

in which $n_\psi$ denotes the number of material phases related to the $\psi$th component and $f_\psi$ defines the composition function of related material phases.

### 5.5.2. *Sensitivity analysis and examples with implicit functions*

The sensitivity analysis with respect to the pseudo-density variables is similar to the standard topology optimization and will not be provided here. Suppose $s_\psi$ represents one of the geometry design variables $x_\psi$, $y_\psi$, $\theta_\psi$ of the $\psi$th component. The derivative of the element elastic modulus with respect to $s_\psi$ is

$$\begin{aligned}\frac{\partial E_i}{\partial s_\psi} &= (\eta_i)^p \frac{\partial}{\partial \xi_\psi}\left( E_0 + \sum_{\varepsilon=1}^{n_c}\sum_{\varsigma=1}^{n_\psi} \mathrm{A}(\Phi_{\psi\varsigma}(\mathbf{x}_i,\mathbf{S}_\psi)) f_\psi\left(E_0, E_{\psi 1},\dots,E_{\psi n_\psi}\right)\right) \\ &= (\eta_i)^p \sum_{\varsigma=1}^{n_\psi} \frac{\partial \mathrm{A}(\Phi_{\psi\varsigma}(\mathbf{x}_i,\mathbf{S}_\psi))}{\partial s_\psi} f_\psi\left(E_0, E_{\psi 1},\dots,E_{\psi n_\psi}\right)\end{aligned} \quad [5.61]$$

By applying the chain rule, the above equation can be further expressed as:

$$\frac{\partial E_i}{\partial s_\psi} = (\eta_i)^p \sum_{\varsigma=1}^{n_\psi} \frac{\partial \mathrm{A}(\Phi_{\psi\varsigma}(\mathbf{x}_i,\mathbf{S}_\psi))}{\partial \Phi_{\psi\varsigma}(\mathbf{x}_i,\mathbf{S}_\psi)} \frac{\partial \Phi_{\psi\varsigma}(\mathbf{x}_i,\mathbf{S}_\psi)}{\partial s_\psi} f_\psi\left(E_0, E_{\psi 1},\cdots,E_{\psi n_\psi}\right) \quad [5.62]$$

Submitting the derivative of the compressing function, we finally have:

$$\frac{\partial E_i}{\partial s_\psi} = (\eta_i)^p \sum_{\varsigma=1}^{n_\psi} \frac{f_\psi\left(E_0, E_{\psi 1},\dots,E_{\psi n_\psi}\right)}{\pi\left(1+\left(\Phi_{\psi\varsigma}(\mathbf{x}_i)/\Delta\right)^2\right)} \frac{\partial \Phi_{\psi\varsigma}(\mathbf{x}_i,\mathbf{S}_\psi)}{\partial s_\psi} \quad [5.63]$$

For a certain component with its implicit function, i.e. $\Phi_\psi(\mathbf{x}_i,\mathbf{S}_\psi)$, its derivative can be analytically calculated according to the explicit relationship between the function and the component location design variables. For example, if the $\psi$th component is a circle, the higher dimensional function used for implicit representation can be

$$\Phi_\psi(\mathbf{x}_\psi, x_\psi, y_\psi) = R_\psi^2 - (x_i - x_\psi)^2 - (y_i - y_\psi)^2 \quad [5.64]$$

Thus, the derivatives of the above function with respect to $x_\psi$ and $y_\psi$ are:

$$\begin{cases} \dfrac{\partial \Phi_\psi(\mathbf{x}_\psi, x_\psi, y_\psi)}{\partial x_\psi} = 2\left(x_i - x_\psi\right) \\ \dfrac{\partial \Phi_\psi(\mathbf{x}_\psi, x_\psi, y_\psi)}{\partial y_\psi} = 2\left(y_i - y_\psi\right) \end{cases} \quad [5.65]$$

If the $\psi$th component is a rectangle, the higher dimensional function used for representation is composed with two subfunctions as follows:

$$\begin{cases} \Phi_\psi(\mathbf{x}_i, x_\psi, y_\psi, \theta_\psi) = f_1(\mathbf{x}_i) + f_2(\mathbf{x}_i) - \sqrt{f_1^2(\mathbf{x}_i) + f_2^2(\mathbf{x}_i)} \\ f_1(\mathbf{x}_i) = \mathrm{b}_\psi^2 - \left(y_\psi^*\right)^2 ; \; f_2(\mathbf{x}_i) = \mathrm{a}_\varepsilon^2 - \left(x_\psi^*\right)^2 \end{cases} \quad [5.66]$$

Note that rotational location design variables need to be considered for a rectangular component. $x^*_\psi$ and $y^*_\psi$ are transformed variables of $x_\psi$ and $y_\psi$ with respect to $\theta_\psi$.

$$\begin{Bmatrix} x_\psi^* \\ y_\psi^* \end{Bmatrix} = \begin{bmatrix} \cos\theta_\psi & -\sin\theta_\psi \\ \sin\theta_\psi & \cos\theta_\psi \end{bmatrix} \begin{Bmatrix} x_i - x_\psi \\ y_i - y_\psi \end{Bmatrix} \quad [5.67]$$

Thus, the derivative can be calculated by the chain rule as follows:

$$\frac{\partial \Phi_\psi(\mathbf{x}_i, x_\psi, y_\psi, \Theta_\psi)}{\partial s_\psi} = \frac{\partial f_1}{\partial s_\psi} + \frac{\partial f_2}{\partial s_\psi} - \frac{1}{\sqrt{f_1^2 + f_2^2}} \left( f_1 \frac{\partial f_1}{\partial s_\psi} + f_2 \frac{\partial f_2}{\partial s_\psi} \right) \quad [5.68]$$

The derivatives of the two subfunctions are

$$\frac{\partial f_1}{\partial s_\psi} = \begin{cases} 2y_{\psi}^*\sin\theta_{\psi} \, , \; s_{\psi} = \mathrm{x}_{\psi} \\ 2y_{\psi}^*\cos\theta_{\psi} \, , \; s_{\psi} = \mathrm{y}_{\psi} \\ 2y_\psi^* \left( \left(y_i - y_\psi\right)\sin\theta_\psi - \left(x_i - x_\psi\right)\cos\theta_\psi \right), s_\psi - \theta_\psi \end{cases}$$

$$\frac{\partial f_2}{\partial s_\psi} = \begin{cases} 2x_\psi^* \cos\theta_\psi , s_\psi = x_\psi \\ -2x_\psi^* \sin\theta_\psi , s_\psi = y_\psi \\ 2x_\psi^* \left( \left( x_i - x_\psi \right) \sin\theta_\psi + \left( y_i - y_\psi \right) \cos\theta_\psi \right), s_\psi = \theta_\psi \end{cases} \quad [5.69]$$

A variety of numerical tests are solved here to illustrate the effectiveness and flexibility of the proposed implicit model. A 100 mm × 200 mm rectangular design domain is discretized with 100 × 200 quadrangular plane stress elements, as shown in Figure 5.37(a). The Young's modulus of structural elements is 1 MPa, and the Poisson's ratio is 0.3. A 1 N force is applied on the middle point of the right edge and the left edge is fixed. A pure topology optimization without component is first tested for the purpose of comparison, and the solution is given in Figure 5.37(b) with a 50% volume fraction.

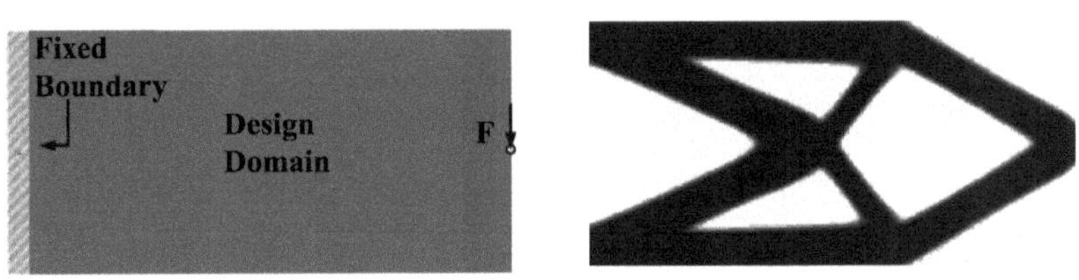


(a) Design domain (b) Topology optimization result

**Figure 5.37.** *The definition of the design domain and a standard topology optimization result*

Then different variants of void components are tested under the same conditions. If we introduce three circular holes, an extremely small value of Young's modulus is thus attributed to the elements inside the hole. Because the initial configuration is symmetrical, the hole is placed in an asymmetric location to generate an original perturbation in sensitivity analysis.

(a) initial state

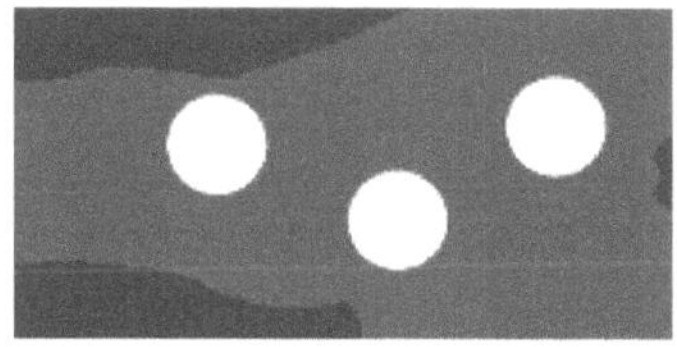

(b) iteration 3

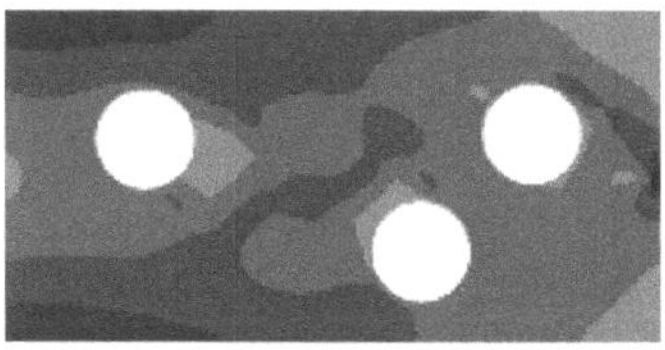

(c) iteration 5

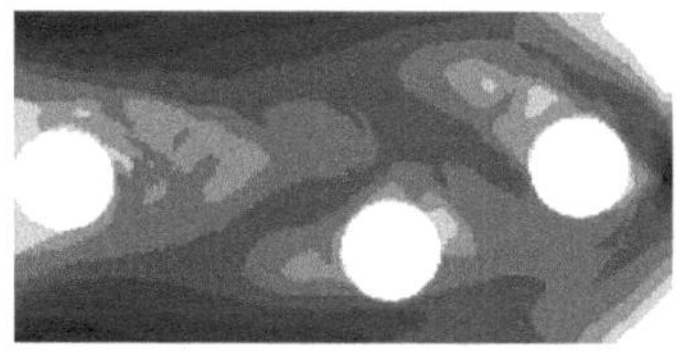

(d) iteration 7

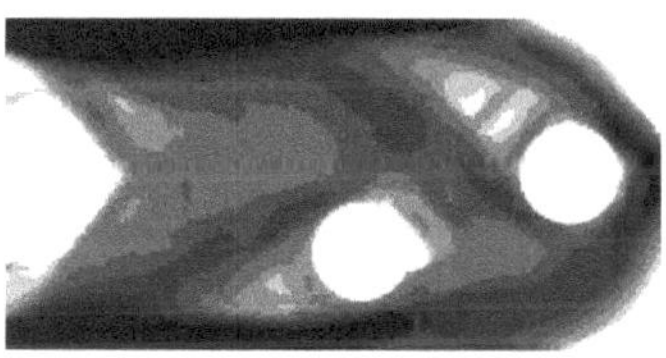

(e) iteration 12

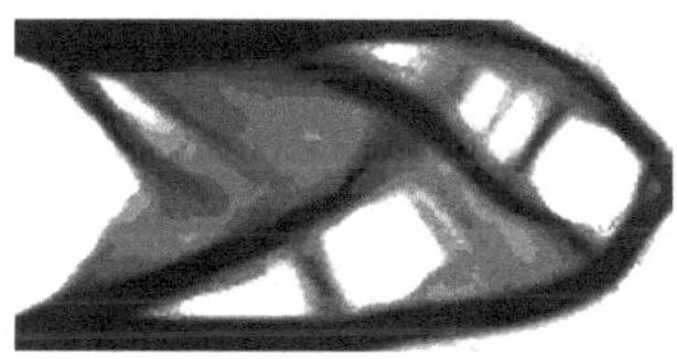

(f) iteration 20

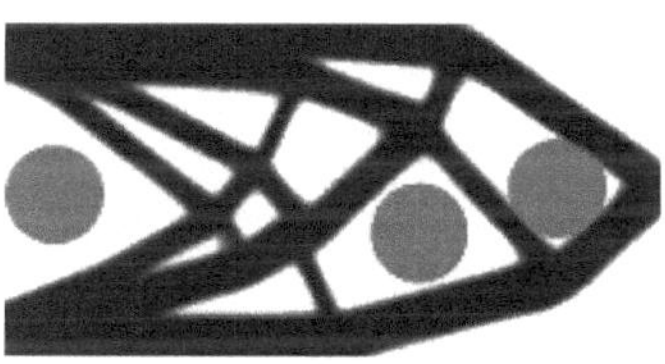

(g) final result

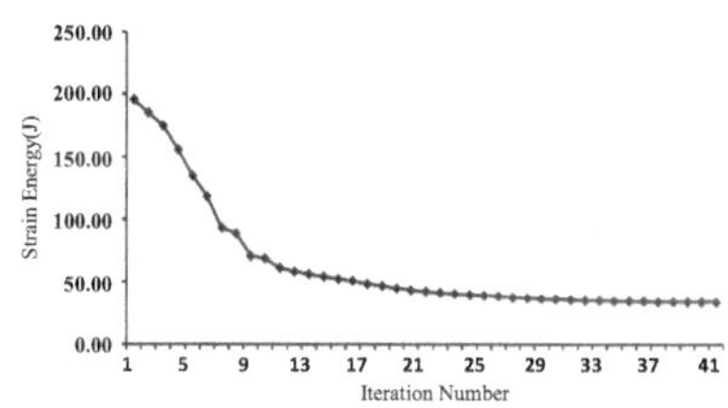


(h) convergent history of the objective function

**Figure 5.38.** *Integrated optimization with three holes*

The three holes are initially located at the coordinates (50 and 70 mm), (100and 30 mm) and (150 and 70 mm). As shown in Figure 5.38, two of the three holes move quickly toward both ends of the design domain and the middle one finds its position in the right part. The existence of the holes does not break the integrity of the structure, but slightly changes the material distribution of the structure.

In some cases, it is necessary to keep the shapes of the holes as functional parts. Therefore, tolerance zones surrounding the holes makes sense in design. Two-phase circular components are defined using R-function. In contrast from the above case, the circular component possesses a certain volume. Therefore, we restrict the volume of the structure together with the component. Figure 5.39 gives the design evolution process and the convergence history. All components are embedded as basic loading parts of the integrated structure due to their stiffness.

(a) initial state

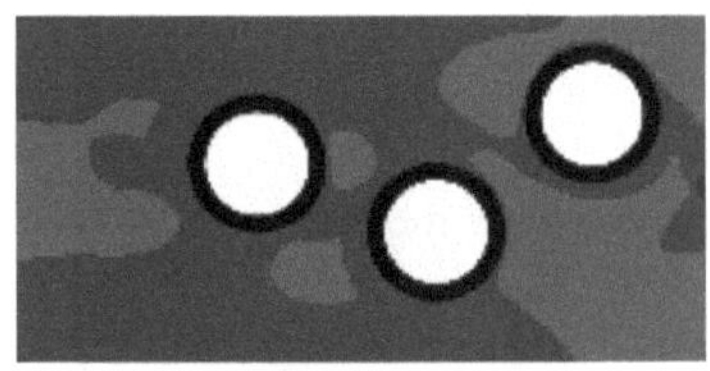

(b) iteration 5

(c) iteration 10

(d) iteration 22

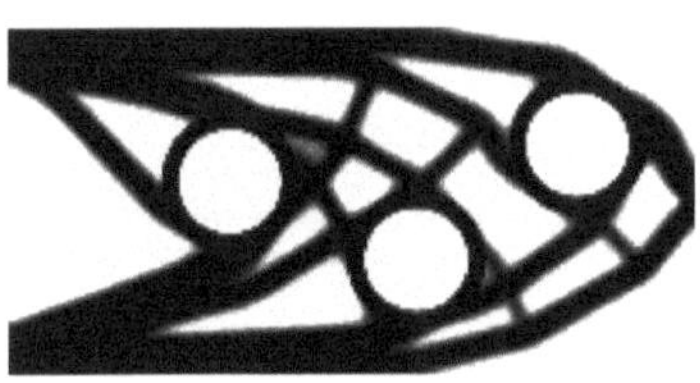

(e) final result

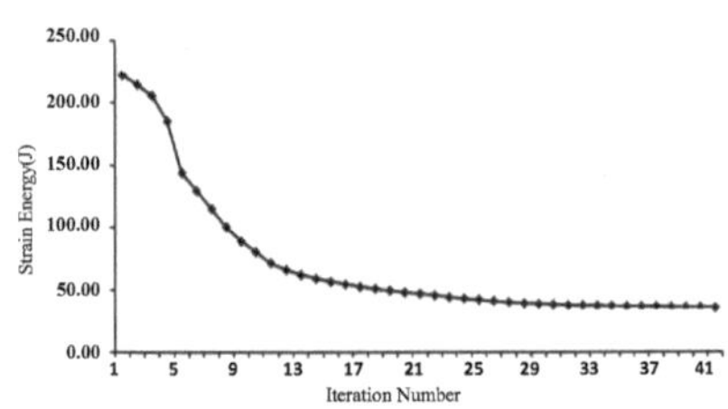


(f) objective convergent history

**Figure 5.39.** *Integrated optimization with three holes with tolerance zones*

The proposed model can easily be extended to three-dimensional problems. On the one hand, a fixed FE mesh greatly simplifies the modeling process. On the other hand, the implicit representation of components greatly saves computational cost in sensitivity analysis w.r.t. location variables. Take a cuboid design domain (30 mm × 60 mm × 90 mm for example), discretized with 30 × 60 × 90 uniform eight-node cubic elements. Material property of the structure is the same as the previous example. The left end of the beam is fixed and a force of 1 N is applied on the bottom edge of the right end. Two solid sphere components with radius 10 mm and Young's modulus 2 Mpa are initially located at the coordinates (15, 30 and 30 mm) and (15, 60 and 30 mm). The volume constraint is set as 18.5%, including the volume occupied by the sphere component. Similarly to the 2D situation, the sphere components move quickly to the fixed end and the loading end due to its high modulus. The optimization evolution is shown in Figure 5.40.

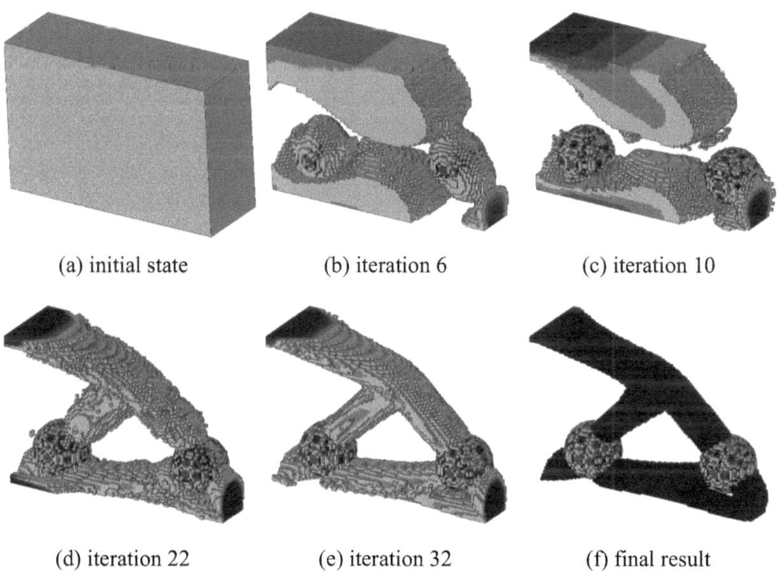

(a) initial state (b) iteration 6 (c) iteration 10

(d) iteration 22 (e) iteration 32 (f) final result

**Figure 5.40.** *Integrated optimization with two sphere components. For a color version of this figure, see www.iste.co.uk/zhang/topology.zip*

### 5.5.3. *Integrated optimization based on XFEM*

The integrated layout and topology optimization of multicomponent systems are further addressed within the XFEM framework here. XFEM using Eulerian mesh is based on a fixed mesh that works with the level set method, which was first proposed by Osher and Sethian [OSH 88], to represent moving interfaces and has been extensively applied to structural optimization [ALL 04b, WAN 03, LUO 07, LUO 08]. XFEM has the advantage of handling problems with material discontinuities across elements. As shown in Figure 5.41, a material interface exists between the structure and embedded component when they have different material attributes. In order to model the local discontinuity, the standard finite element approximation within a narrow domain $\Omega_{enr}$ is enriched in XFEM. Usually, the interface boundary shape is described implicitly as a curve of a higher dimension level set function.

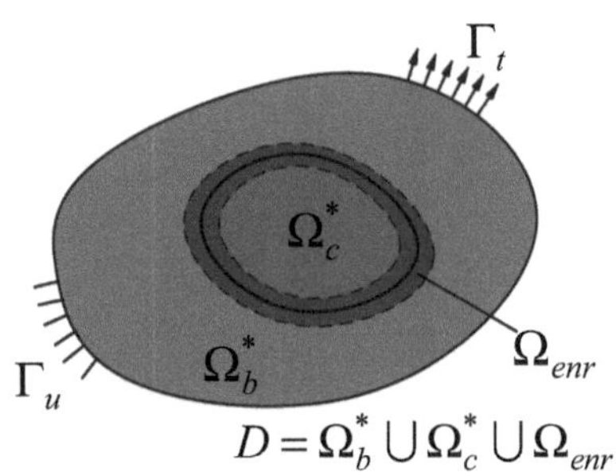


**Figure 5.41.** *Local enrichment of the material interface with XFEM*

To favor the XFEM-based integrated optimization, a general formulation of the strain energy is stated as

$$C(\mathbf{U}) = \int_D \frac{1}{2} \boldsymbol{\varepsilon}^{\mathrm{T}}(\mathbf{U}) \cdot \mathbf{D} \cdot \boldsymbol{\varepsilon}(\mathbf{U}) d\Omega \qquad [5.70]$$

where **D** is the material elastic matrix. $\boldsymbol{\varepsilon}$ is the strain vector. In the case of material–void interface problems (e.g. a hole in the structure), XFEM is always implemented by neglecting the integration over the void part [WEI 10, VAN 07], such as

$$C(\mathbf{U}) = \int_D \frac{1}{2} \boldsymbol{\varepsilon}^{\mathrm{T}}(\mathbf{U}) \cdot \mathbf{D} \cdot \boldsymbol{\varepsilon}(\mathbf{U}) \cdot H(\Phi) d\Omega \qquad [5.71]$$

But for a material–material interface problem (e.g. solid component and supporting structure), additional degrees of freedom have to be introduced in the enriched domain. As a result, the global strain energy can be calculated as:

$$C(\mathbf{U})=\int_{\Omega_b^*}\frac{1}{2}\boldsymbol{\varepsilon}^{\mathrm{T}}(\mathbf{U})\cdot\mathbf{D}_b\cdot\boldsymbol{\varepsilon}(\mathbf{U})d\Omega+\int_{\Omega_c^*}\frac{1}{2}\boldsymbol{\varepsilon}^{\mathrm{T}}(\mathbf{U})\cdot\mathbf{D}_c\cdot\boldsymbol{\varepsilon}(\mathbf{U})d\Omega$$
$$+\int_{\Omega_{enr}}\frac{1}{2}\boldsymbol{\varepsilon}^{\mathrm{T}}(\mathbf{U},\Phi)\cdot\mathbf{D}(\Phi)\cdot\boldsymbol{\varepsilon}(\mathbf{U},\Phi)d\Omega \quad [5.72]$$

Here, the level set function is chosen as the signed distance function with the definition of

$$\Phi(\mathbf{x},t)=\pm\min_{\mathbf{x}_\Gamma\in\Gamma(t)}\left\|\mathbf{x}-\mathbf{x}_\Gamma\right\| \quad [5.73]$$

The sign is positive or negative when $\mathbf{x}$ is outside or inside the design domain boundary $\Gamma(t)$. $t$ is a pseudo-time describing the shape variation process of a component in shape sensitivity analysis.

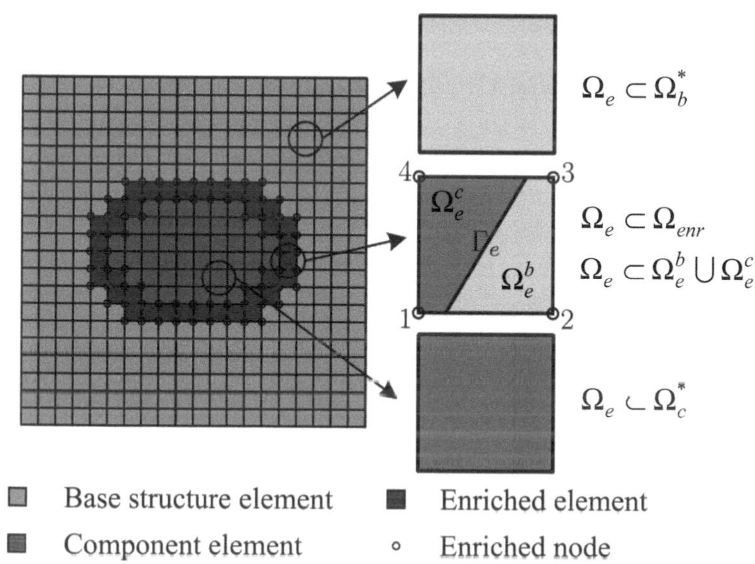


**Figure 5.42.** *Partition of structural region and element classifications*

Practically, the base domain is uniformly discretized with finite element meshes. Figure 5.42 shows the structural elements, component elements and enriched elements. The last set belongs to bimaterial elements and others are one-material elements. A bi-material element domain can be considered as a combination of the structure part and the component part.

In XFEM, the level set surface is always approximated by discrete values at nodes and then interpolated over the elements by finite element shape functions as:

$$\Phi^h(\mathbf{x}) = \sum_{i \in I} \mathbf{N}_i(\mathbf{x})\Phi_i \qquad [5.74]$$

where $\Phi^h(\mathbf{x})$ refers to the approximation of the level set surface. $\mathbf{N}_i$ denotes the standard finite element shape functions. $I$ represents the set of all nodes in the domain.

The material elastic matrix at a spatial point $\mathbf{x}$ within the structure can thus be expressed by the level set function $\Phi^h(\mathbf{x})$ and the Heaviside function $H()$.

$$\mathbf{D}(\mathbf{x}) = \mathbf{D}_s + (\mathbf{D}_c - \mathbf{D}_s)H\left(\Phi^h(\mathbf{x})\right) \qquad [5.75]$$

where $\mathbf{D}_s$ and $\mathbf{D}_c$ are the elastic matrices related to the structure and component, respectively. The above relationship means that any point $\mathbf{x}$ in the component domain ($\Phi^h(\mathbf{x}) > 0$) will have an elastic matrix equal to $\mathbf{D}_c$. Otherwise, the elastic matrix will be equal to $\mathbf{D}_s$. Once the coordinates of a spatial point in the structure are given, the material property can be determined.

Under the XFEM framework, the displacement field is interpolated by

$$\mathbf{U}^h(\mathbf{x}) = \sum_{i \in I} \mathbf{N}_i \mathbf{U}_i + \sum_{i \in I^*} \mathbf{N}_i \varphi \mathbf{a}_i \qquad [5.76]$$

where $\mathbf{U}_i$ is the nodal displacement vector of the standard FE part. $\mathbf{a}_i$ is the additional nodal unknown vector. $I^*$ is the set of all enriched nodes, which is a subset of $I$. Function $\varphi$ is the enrichment function whose selection depends upon the nature of the problem.

For the integrated optimization with solid components, which is considered a weak discontinuity, the following function proposed by Moës *et al.* [MOE 03] can be used as the enrichment function:

$$\varphi(\mathbf{x})=\sum_{i\in I}\mathbf{N}_i(\mathbf{x})|\Phi_i| - \left|\sum_{i\in I}\mathbf{N}_i(\mathbf{x})\Phi_i\right| \quad [5.77]$$

In the case of material–void interface problems, there are no additional degrees of freedom, a change will be introduced in the approximation of displacement so that:

$$\mathbf{U}^h(\mathbf{x})=\sum_{i\in I}\mathbf{N}_i(\mathbf{x})H(\mathbf{x})\mathbf{U}_i \quad [5.78]$$

Here, an analytical XFEM-based sensitivity calculation method is developed. The derivative of an element stiffness matrix $\mathbf{K}_e$ can be calculated via the chain rule:

$$\frac{\partial \mathbf{K}_e}{\partial s_\psi}=\sum_{i\in I}\frac{\partial \mathbf{K}_e}{\partial \Phi_i}\frac{\partial \Phi_i}{\partial s_\psi} \quad [5.79]$$

The derivative of the structural strain energy with respect to nodal level set function value $\Phi_i$ can be written as:

$$\frac{\partial C}{\partial \Phi_i}=\sum_{e\in\Omega_{enr}}\left(-\frac{1}{2}\mathbf{U}_e^{\mathrm{T}}\cdot\int_{\Omega_e}\left(\mathbf{B}^{\mathrm{T}}\mathbf{D}\frac{\partial \mathbf{B}}{\partial \Phi_i}+\frac{\partial \mathbf{B}^{\mathrm{T}}}{\partial \Phi_i}\mathbf{D}\mathbf{B}+\mathbf{B}^{\mathrm{T}}\frac{\partial \mathbf{D}}{\partial \Phi_i}\mathbf{B}\right)d\Omega\cdot\mathbf{U}_e\right) \quad [5.80]$$

Due to the fact that any change of nodal level set function value $\Phi_i$ for a one-material element does not affect element stiffness matrices, it follows that:

$$\frac{\partial C}{\partial \Phi_i}=\sum_{e\in\Omega_{enr}}\left(-\frac{1}{2}\mathbf{U}_e^{\mathrm{T}}\cdot\frac{\partial \mathbf{K}_e}{\partial \Phi_i}\mathbf{U}_e\right) \quad [5.81]$$

Comparing the above two equations, the derivative of element stiffness matrix $\mathbf{K}_e$ can be written as:

$$\frac{\partial \mathbf{K}_e}{\partial \Phi_i} = \int_{\Omega_e} \left( \mathbf{B}^{\mathrm{T}} \mathbf{D} \frac{\partial \mathbf{B}}{\partial \Phi_i} + \frac{\partial \mathbf{B}^{\mathrm{T}}}{\partial \Phi_i} \mathbf{D} \mathbf{B} + \mathbf{B}^{\mathrm{T}} \frac{\partial \mathbf{D}}{\partial \Phi_i} \mathbf{B} \right) d\Omega \qquad [5.82]$$

The derivative of the geometry matrix **B** can be derived according to the discretization of the level set functions using the shape functions. The derivatives of the level set value with respect to the geometry variables can then be approximated by finite difference as:

$$\frac{\partial \Phi_i}{\partial s_\psi} \approx \frac{\Phi_i \left( s_\psi + \Delta s_\psi \right) - \Phi_i \left( s_\psi \right)}{\Delta s_\psi} \qquad [5.83]$$

Numerical examples with different component geometries are studied to verify the proposed approach. Here, the elastic modulus of the design domain and the components are $7 \times 10^{10}$ Pa and $2 \times 10^{11}$ Pa, respectively. The strain energy of the global structure is minimized subject to the non-overlapping constraints based on FCM. Besides, the volume fraction is constrained to 35% of the whole domain.

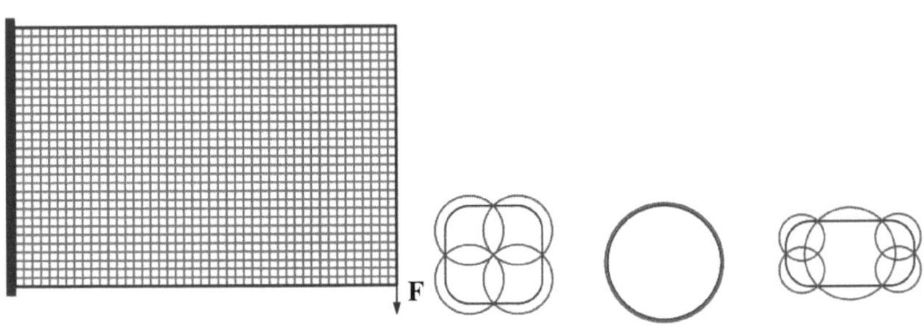


**Figure 5.43.** *Definition of the design domain and the finite circles for the components*

Figure 5.43 shows a 45 m × 30 m design domain discretized with 45 × 30 four nodes quadrilateral finite elements. A point force of 100,000 N is applied on the lower left corner. To avoid possible overlap between components, each of them is approximated with a specific number of circles, as illustrated also in Figure 5.43. Iterative design patterns are shown in Figure 5.44. In the final design, three components act as essential parts of the structure due to their high elastic modulus.

The convergence histories of the objective function and the volume fraction are also shown in Figure 5.44. It can be seen that the volume fraction reaches its upper bound. There also exists an oscillation of the volume fraction, caused mainly by the movements of components.

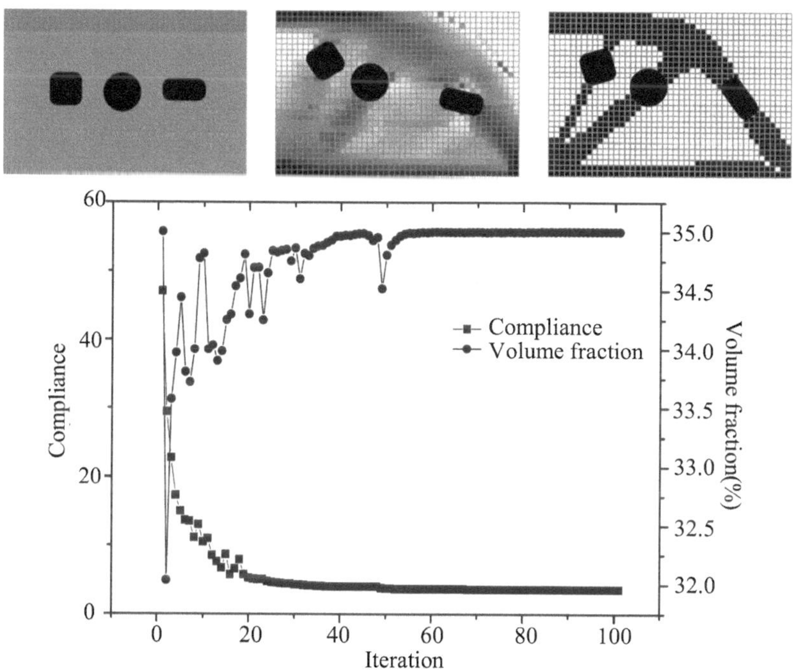


**Figure 5.44.** *The convergence history of the structural configuration, objective function and the structural volume*

Another example is to include two L-shaped solid components. The structural domain and boundary conditions are the same as in the previous example. As shown in Figure 5.45, each component is approximated with nine finite circles. Initial values of pseudo-densities are set at 0.35. Iterative design patterns are shown in

Figure 5.46. Both components are properly placed as essential parts of the structure. The iteration history of the objective function has a stable convergence, as shown in Figure 5.46.

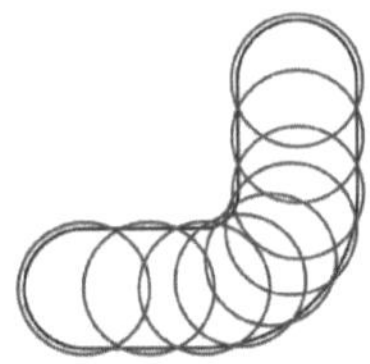

**Figure 5.45.** *Finite circles approximation of an L-shaped component*

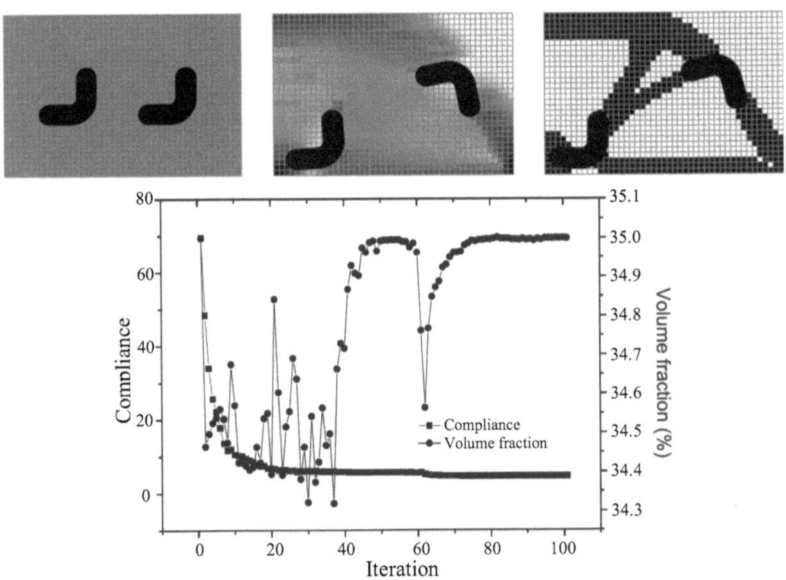


**Figure 5.46.** *The convergence history of the structural configuration, objective function and the volume fraction*

## 5.6. Conclusions

Several important issues such as non-overlapping constraints and integrated modeling for integrated layout and topology optimization are discussed in this chapter. First, based on the finite circle method, proper circle definitions are used to approximate the contour of the components and to formulate the non-overlapping constraints. To avoid using large numbers of non-overlapping constraints, an improved constraint aggregation method based on the KS function and a Steffensen iteration is proposed to aggregate the constraints into a single constraint. With an additional complex step derivative approximation, the design sensitivities can be precisely obtained.

Second, based on body-fit modeling, density points and embedded meshing techniques are proposed to establish exact nodal connections between the components and to ensure the simultaneous optimization of different design variables. Instead of using global element remeshing, local embedded meshing and superelement techniques are used to improve the efficiency of the modeling process. The MPC connections between components and structures are later introduced to simulate bolt or rivet joints.

The implicit model for the component is then introduced. The contour of the component is described by level set function. Elements located on the boundary of the component will thus have their material properties interpolated or described by XFEM. The movement of a component is actually represented by a movement of the component material property, without any element remesh. Benefitting from the fixed finite element, the design sensitivities with respect to the geometry design variables can be transferred into the derivatives of the level set functions and those with respect to the pseudo-densities in turn. The optimization iteration will thus avoid the procedures of finite difference or semi-analytical calculation with additional finite element analysis.

Using the above techniques, integrated layout and topology optimization can be implemented in different numerical examples. The convergence of the objective functions, the effect of different optimization strategies, and the applications are discussed in detail with reasonable results obtained.

# 6

# Optimization with Constraints on Multifastener Joint Loads

## 6.1. Introduction

In an assembled aircraft structure, bolts or rivets are widely used as multifastener joints. They are sometimes the weakest component of a structure due to the high intensity of joint load [NIU 88, BAR 92, WAN 00, CHI 10]. Earlier studies were focused on developing analytical and numerical methods for stress and failure predictions of multifastener joints. Typical models concerned panels joined by single or multiple joints, in which joint loads as well as stress distributions around pin holes, etc., are mostly analyzed [ROW 82, WAN 88, ZHA 96, CAM 97]. Poon and Xiong [POO 95] and Oh *et al.* [OH 97] considered the optimization of fastener joint locations, ply angles and stacking sequences of laminates, fastener diameters and edge distances, etc., to avoid the failure of fasteners. Bianchi *et al.* [BIA 07] developed an optimization procedure maximizing the load-carrying capability of the joint system to balance the number and size of bolts. Ekh and Schön [EKH 08] evaluated the effects of different parameters on the load distribution, such as the mismatch of member plates, length of the overlap region and the fastener's stiffness. Optimization was then carried out to minimize the bearing stress. In the work of Oinonen *et al.* [OIN 10], a "weakest link" method was proposed to optimize the layout of fasteners for the bracket-to-beam joints. The design objective was to ameliorate Von-Mises equivalent strain as well as shear loads in the joints.

Attempts were also made by Jiang and Chirehdast [JIA 97], Buhl [BUH 01], Zhu and Zhang [ZHU 06a], Qiao and Liu [QIA 09], and Zhu and Zhang [ZHU 10a] to introduce the concept of topology optimization into the layout design of fixations and fasteners. In the works of Chickermane and Gea [CHI 97], Li *et al.* [LI 01] and Qian and Ananthasuresh [QIA 04], structural patterns of different parts and the layout of fasteners were also sought by constraining these subdomains with different volume fractions. Chickermane *et al.* [CHI 99] also proposed a topology optimization method for the location optimization of fasteners in conjunction with fastener load constraints, while the connected components remain unchanged.

Unfortunately, it was found in engineering practices that the adjustment of joint distributions, the addition of more joints or the enlargement of joint diameters might not lower the joint loads significantly due to the stiffness mismatch between the connected structure components. This is especially the case for aircraft spar-skin structure with overloaded fastener joints. For example, during a typical aircraft wing beam design, the joint loads in the fasteners near the root of the beam were found to be unexpectedly large. To avoid failure of structural joints, possible solutions include increasing diameters of the fasteners and connected spars, rearranging the layout of the fasteners or even adding more fasteners, as shown in Figure 6.1, which will unfortunately increase the weight of the structure.

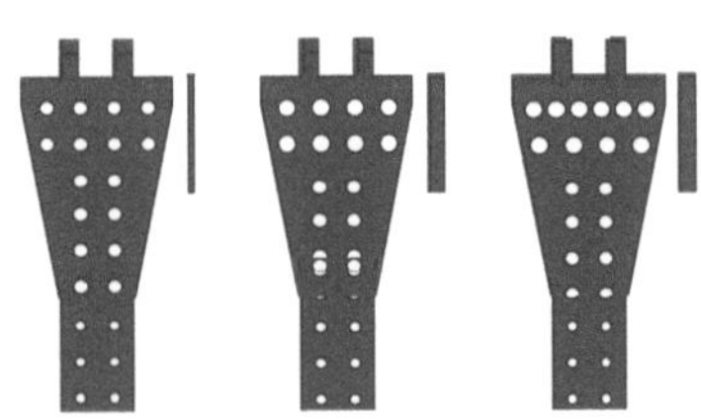

**Figure 6.1.** *Existing solutions to avoid the failure of fasteners*

In fact, joint loads highly depend not only on mechanical and geometrical attributes of joints themselves, but also on stiffness distributions of the connected structures according to aircraft design manuals [NIU 88, WAN 00] and existing works [POO 95, OH 97, EKH 08, CHI 10]. In this chapter, the standard topology optimization method is applied to deal with the above problem by introducing additional multifastener joint load constraints for the first time. Unlike the existing works, dominant shear loads over joints are controlled by optimizing the stiffness distribution of connected structures. In the calculation of shear loads, short beam elements are used to model multifastener joints with the negligence of bolt-hole clearance, clamp-up and friction effects. The proposed optimization method can be considered an effective way to limit the joint loads during the design procedure of assembled aircraft structures. Meanwhile, designers can also benefit from the optimized load carrying path for detailed structure design.

## 6.2. Joint load calculation and sensitivity analysis

In the topology optimization model, a precise calculation of shear loads in fasteners can involve deep studies of relative displacements of hole centers, fastener deformation, conforming contact between fasteners and member plates as well as effects of fastener clamp-up, friction and out of plane deformation, etc. This brings extra complexities and intractable computational difficulties. Here, multifastener joints are approximated by numbers of beam elements as used by Chickermane *et al.* [CHI 99], Ekh and Schön [EKH 08].

For a short beam of two nodes A and B shown in Figure 6.2, the total shear load is calculated as:

$$F_j = \sqrt{F_{jx}^2 + F_{jy}^2} \qquad [6.1]$$

with:

$$\begin{aligned}
F_{jx} &= \frac{12EI}{L^3(1+\Theta)} \cdot u_{jAx} - \frac{12EI}{L^3(1+\Theta)} \cdot u_{jBx} + \frac{6EI}{L^2(1+\Theta)} \cdot \theta_{jAy} + \frac{6EI}{L^2(1+\Theta)} \cdot \theta_{jBy} \\
F_{jy} &= \frac{12EI}{L^3(1+\Theta)} u_{jAy} \quad \frac{12EI}{L^3(1+\Theta)} \cdot u_{jBy} - \frac{6EI}{L^2(1+\Theta)} \cdot \theta_{jAx} - \frac{6EI}{L^2(1+\Theta)} \cdot \theta_{jBx} \qquad [6.2] \\
\Theta &= \frac{12EI\gamma}{GSL^2}
\end{aligned}$$

where $E$ and $G$ are the elastic and shear moduli. $I$, $S$ and $L$ denote the moment of inertia, the cross-section area and the beam length. $\Theta$ is the shear coefficient. $\gamma = 10/9$ is the shear factor of the cross-section. $u_{jAx}$ and $\theta_{jAx}$ are the corresponding nodal displacement and rotation angle of node $A$ in $x$ direction. Similar definitions are used for $u_{jAy}$, $\theta_{jAy}$, $u_{jBx}$, $\theta_{jBx}$, $u_{jBy}$ and $\theta_{jBy}$.

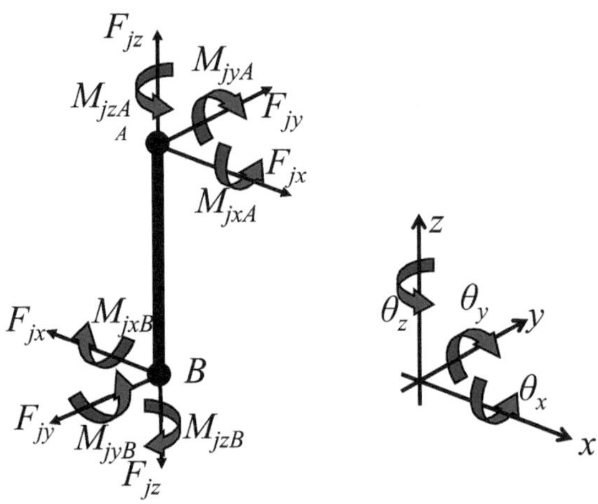


**Figure 6.2.** *Beam element and its coordinate system, tension force, moments and shear forces*

Sensitivity analysis of elastic strain energy with respect to pseudo-densities is very popular. Here, we are focused on the sensitivity analysis of the joint load $F_j$, whose components are denoted by $F_{jx}$ and $F_{jy}$. For example, $F_{jx}$ can be expressed as a linear function of nodal displacements

$$F_{jx} = \frac{12EI}{L^3(1+\Theta)} \cdot u_{jAx} - \frac{12EI}{L^3(1+\Theta)} \cdot u_{jBx} + \frac{6EI}{L^2(1+\Theta)} \cdot \theta_{jAy} + \frac{6EI}{L^2(1+\Theta)} \cdot \theta_{jBy} \qquad [6.3]$$
$$= \boldsymbol{\lambda}_{jx}^T \mathbf{U}$$

where $\boldsymbol{\lambda}_{jx}$ is a constant vector with the same dimension as the displacement vector $\mathbf{U}$. Items of $\boldsymbol{\lambda}_{jx}$ corresponding to the four degrees of freedom are defined by the coefficients in the above equation, while the rest are set to be zero. Accordingly, the sensitivity of $F_{jx}$ is calculated as:

$$\frac{\partial F_{jx}}{\partial \eta_i} = \frac{\partial\left(\boldsymbol{\lambda}_{jx}^T \mathbf{U}\right)}{\partial \eta_i} = \boldsymbol{\lambda}_{jx}^T \cdot \frac{\partial \mathbf{U}}{\partial \eta_i} \quad [6.4]$$

Similar calculations of $F_{jy}$ can be performed to obtain constant vector $\boldsymbol{\lambda}_{jy}$. As a result, the derivative of the total joint load corresponds to:

$$\begin{aligned}\frac{\partial F_j}{\partial \eta_i} &= \frac{\partial\left(\sqrt{F_{jx}^2 + F_{jy}^2}\right)}{\partial \eta_i} = \frac{1}{2\sqrt{F_{jx}^2 + F_{jy}^2}} \cdot \frac{\partial\left(F_{jx}^2 + F_{jy}^2\right)}{\partial \eta_i} \\ &= \frac{1}{2F_j} \cdot \left(2F_{jx}\boldsymbol{\lambda}_{jx}^{\mathrm{T}} \cdot \frac{\partial \mathbf{U}}{\partial \eta_i} + 2F_{jy}\boldsymbol{\lambda}_{jy}^{\mathrm{T}} \cdot \frac{\partial \mathbf{U}}{\partial \eta_i}\right) \\ &= \frac{F_{jx}\boldsymbol{\lambda}_{jx}^{\mathrm{T}} + F_{jy}\boldsymbol{\lambda}_{jy}^{\mathrm{T}}}{F_j} \cdot \frac{\partial \mathbf{U}}{\partial \eta_i} \\ &= \boldsymbol{\lambda}_j^{\mathrm{T}} \cdot \frac{\partial \mathbf{U}}{\partial \eta_i}\end{aligned} \quad [6.5]$$

Based on the finite element equilibrium equation, the above equation can further be written as:

$$\boldsymbol{\lambda}_j^{\mathrm{T}} \cdot \frac{\partial \mathbf{U}}{\partial \eta_i} = \boldsymbol{\lambda}_j^{\mathrm{T}} \cdot \left[-\mathbf{K}^{-1}\left(\frac{\partial \mathbf{F}}{\partial \eta_i} - \frac{\partial \mathbf{K}}{\partial \eta_i}\mathbf{U}\right)\right] = -\left(\mathbf{K}^{-1}\boldsymbol{\lambda}_j\right)^{\mathrm{T}} \cdot \left(\frac{\partial \mathbf{F}}{\partial \eta_i} - \frac{\partial \mathbf{K}}{\partial \eta_i}\mathbf{U}\right) \quad [6.6]$$

The above expression can be simplified by the following notation:

$$\mathbf{K}^{-1}\boldsymbol{\lambda}_j = \mathbf{U}^* \quad [6.7]$$

Notice that the stiffness matrix **K** is symmetric. **U*** can be interpreted as the displacement vector related to the adjoint load vector $\boldsymbol{\lambda}_j$ applied on the structure. The substitution gives rise to:

$$\frac{\partial F_j}{\partial \eta_i} = -\left(\mathbf{U}^*\right)^{\mathrm{T}} \cdot \left(\frac{\partial \mathbf{F}}{\partial \eta_i} - \frac{\partial \mathbf{K}}{\partial \eta_i}\mathbf{U}\right) \quad [6.8]$$

In the above equation, the derivative of the load vector is zero when only design-independent loads are applied to the structure. Derivatives of the stiffness matrix **K** with respect to the pseudo-densities are easily obtained by means of SIMP material interpolation model used in this chapter.

In practice, sensitivity analyses of the joint loads will be computationally expensive due to the additional finite element analyses required for large numbers of fasteners and joint load constraints. To solve the problem, methods like KS function and P-norm function can be used to merge large numbers of design constraints into only one constraint. In addition, as shown in numerical examples, only a small number of fasteners need to be constrained, because the rest joint loads are small enough to be neglected during optimization.

In the following sections, two numerical examples are presented to illustrate the efficiency and validity of the proposed optimization model. One is further validated with a loading test of fabricated resin models. We use the optimization algorithm GCMMA (globally convergent method of moving asymptotes, [SVA 95]) implemented in the general-purpose design platform BOSS Quattro [RAD 02]. The density filter technique [BRU 01] is applied, with the filter radius being three times of the average element size in both examples to avoid the checkerboards.

## 6.3. Numerical examples and discussions

### 6.3.1. *Cantilever beam with experiments*

Here, we will design an I-shaped cantilever beam joined with a thin sheet by two rows of evenly distributed fasteners. Dimensions and layout of joints are shown in Figure 6.3. As the shear load distributions are identical in both rows, fasteners are numbered only for one row from the beam root to the tip.

The assembled structure is meshed with quadrangular shell elements of size 1 mm × 1 mm and thickness 1 mm. The design domain of the structure is the I-shaped beam web. The elastic modulus of the solid material is $2.6 \times 10^9$ Pa and Poisson's ratio is 0.3.

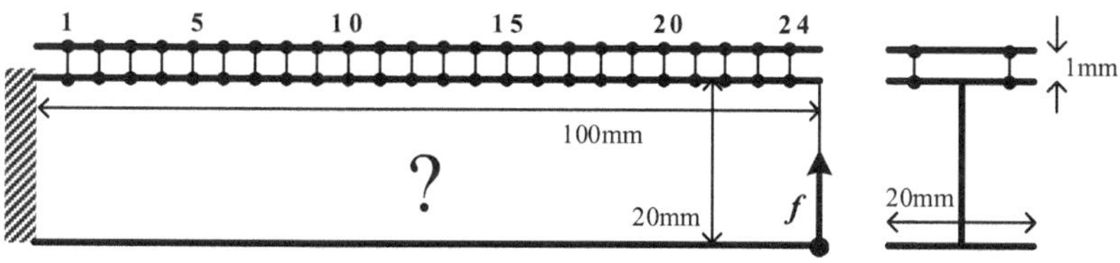


**Figure 6.3.** *I-shaped cantilever beam connected with a thin sheet*

A point force of 150 N is applied on the bottom right corner of the design domain. In Figure 6.4, we present the shear load distribution over fastener joints after static analysis of the initial solid FE model. We find that the maximum shear load occurs at the beam root, which indicates the most vulnerable position.

For the purpose of comparison, a standard strain energy minization topology optimization is firstly carried out without joint load constraints. The only constraint is defined by limiting the volume fraction of the design domain to 30%. Figure 6.5 presents the optimized design and related joint loads after 36 design iterations. The mean compliance is 0.660 J. The maximum tip displacement is 8.887 mm, while the maximum shear load rises to 30.13 N at the beam root. The optimized topology seems to be the classical topology solution of a pure cantilever beam. There are no materials distributed near the joint for the structure reinforcement, even where the maximum shear load occurs, which implies that the increase in the maximum joint load does not significantly affect the minimization of the mean compliance.

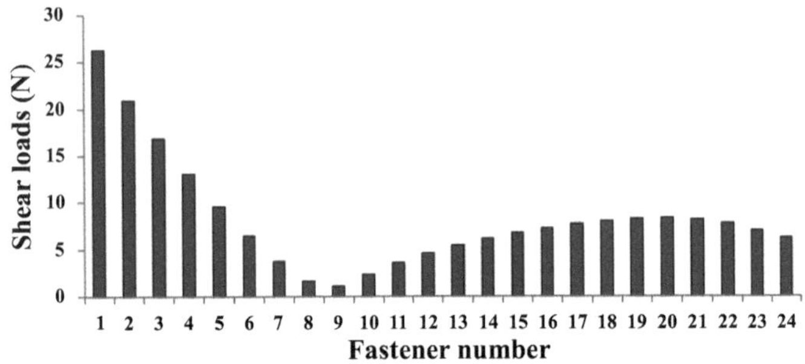


**Figure 6.4.** *The initial distribution of the shear loads in the joints*

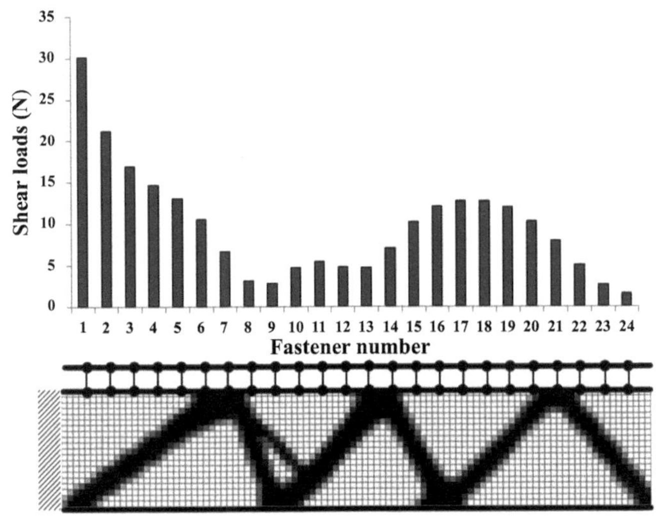


**Figure 6.5.** *Optimized design and joint load distribution without joint load constraint (mean compliance 0.660 J, maximum joint load 30.13 N)*

Now, the same topology optimization problem is considered with the introduction of joint load constraints. The upper bound of the shear load is limited to 20 N. Due to the symmetry, we only constrain the shear loads in one row of fasteners. Figure 6.6 shows the optimized design and joint loads distribution after 30 design iterations. To ease the comparison, load distribution in Figure 6.5 is also plotted here.

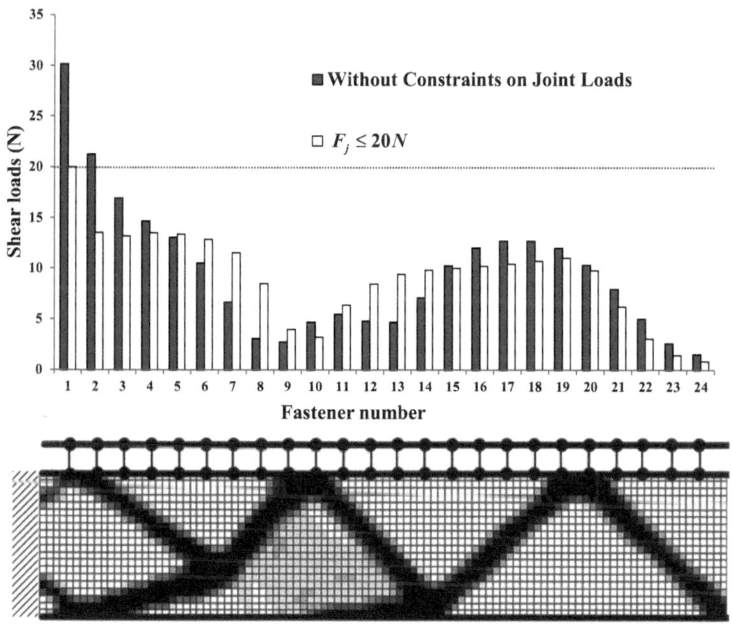


**Figure 6.6.** *Optimized design and comparison of joint load distribution with joint load constraint (Mean compliance 0.667 J, maximum joint load 19.97 N)*

Detailed results are listed and compared in Table 6.1. The mean compliance is now 0.667 J, while the maximum tip displacement reaches 8.988 mm. Although the structural stiffness is slightly smaller than before, the maximum shear load in the joints is now 19.97 N with a great reduction by more than 10 N, i.e. 33.7% of the previous design. The maximum joint load appears still at the beam root and all the loads now satisfy the design constraints.

| Items | Optimized design without joint load constraints | Optimized design with $F_j \leq 20N$ | Initial solid beam |
|---|---|---|---|
| Mean compliance (J) | 0.660 | 0.667 | 0.335 |
| Maximum displacement (mm) | 8.887 | 8.988 | 4.53 |
| Maximum shear loads (N) | 30.13 | 19.97 | 26.3 |
| Volume fractions | 0.3 | 0.3 | 1 |

**Table 6.1.** *Comparison of optimized results*

To have a deep insight into the design solution, the principal stresses near the beam root are plotted and compared in Figure 6.7 for two optimized designs. Notice that the arrows indicate magnitudes and directions of principal stresses. In the optimized design of Figure 6.7, the structural configuration near the beam root is redesigned significantly to provide a reasonable support near the joint where maximum shear loads occur. Compared with the optimized design in Figure 6.5, an additional push is provided by the structure branch to offset the shear effect.

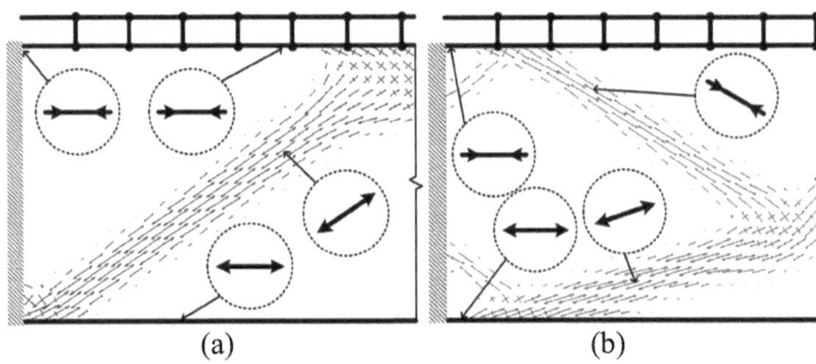


**Figure 6.7.** *Principal stresses of the optimized designs*

To validate the optimized design, loading tests are carried out. We use rapid prototyping to fabricate resin models according to the above two topology optimization solutions. The resin material Somos 14120 is a standard ABS-like stereolithography resin. According to the data obtained from the supplier DSM Somos (www.dsmsomos.com, 14120 white documents) and some previous applications [NOR 09, GU 12], the stereolithography resin material can approximately maintain linear elastic properties within a small deformation.

According to the optimized designs, solid models of engineering features are built in CATIA as shown in Figure 6.8(a) and then outputted as STL models for rapid prototyping. Clearly, structural configurations are globally maintained except for some minor details. As thin-walled stereolithography resin parts have the problem of easy-buckling, some extra stiffeners are added to the structure. Because failures of practical bolt and rivet joints appearing around the joined holes on the member plates are not easily detectable, all the joints are directly fabricated as small vulnerable resin beams of identical size and cross-section with approximately the same strength, as shown in Figure 6.8(b). These resin joints can thus be considered as simple breaking sensors of shear loads. Although the experiment cannot precisely determine the magnitude of the shear loads, a significant tendency should be persuasive due to the fact that the optimized design with joint load constraints remarkably reduces the shear loads.

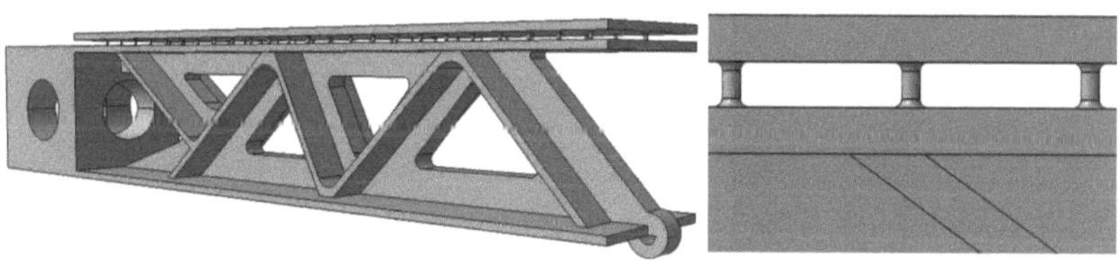

(a) (b)

**Figure 6.8.** *Solid CAD model of the optimized beam and joints with engineering features*

a)

b)

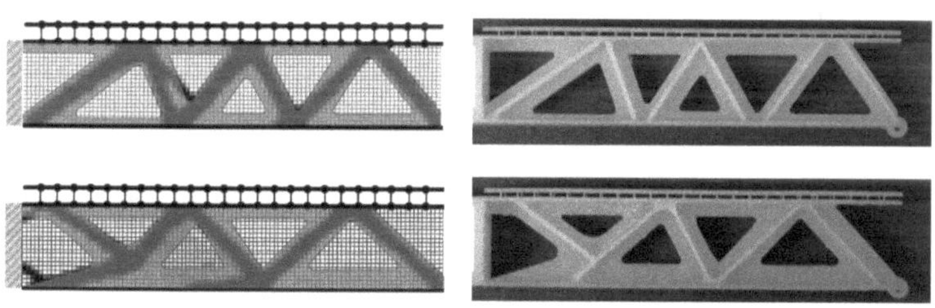

**Figure 6.9.** *Resin models fabricated according to the topology optimization result*

In Figure 6.9, transparent outlines of the resin models are plotted over the material density distribution for comparison. Resin models marked as A and B in Figure 6.9 are then fabricated with a SPS-350B stereolithography machine.

The loading test is set up in the following way: resin models are installed onto static stretching machine as cantilever beams with a fixture as shown in Figure 6.10. A strong string bounds the lower right corner to the dynamometer. The aim of the experiment is to find the minimum stretching force breaking the joints in both models. Three groups of specimens are tested. By slowly increasing the stretching force, the force value is recorded from the dynamometer once the first joint is broken. All the experimental data are compared in Table 6.2, and model B with broken joints is shown in Figure 6.11.

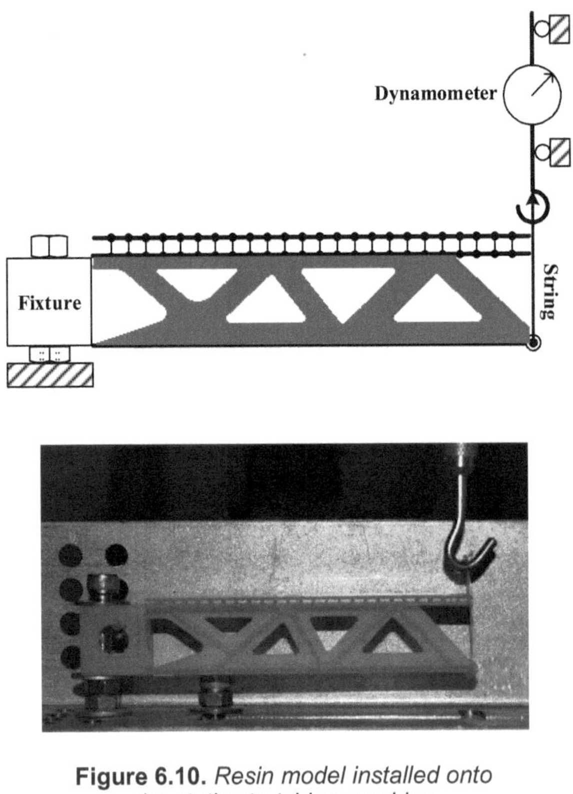


**Figure 6.10.** *Resin model installed onto the static stretching machine*

**Figure 6.11.** *Model B with broken joints*

| | Minimum stretching forces (N) | |
|---|---|---|
| | Model A | Model B |
| Group 1 | 27.6 | 42.8 |
| Group 2 | 32.1 | 40.6 |
| Group 3 | 31.7 | 44.2 |

**Table 6.2.** *Comparison of loading test results*

Magnitudes of the stretching forces in Table 6.2 have clearly indicated that the resin joints related to model B can undertake a larger stretching force. In other words, optimized design with joint load constraints can reduce the shear loads significantly.

It should be noted that the displacements of the resin models are not evaluated in the experiment. This is because the maximum displacements are 0.621 and 0.624 mm, respectively, according to finite element analyses of both models when a 20 N force is applied. The difference is only 0.003 mm, which cannot be precisely distinguished by our existing equipment in the experiment.

Later, the optimization designs with different shear load constraints are implemented and results are achieved as shown in Figure 6.12. We find that shear load constraint affects the optimized configuration significantly. Meanwhile, the strain energy of the optimized structure will be greater than the standard one. As the constraint upper bounds decrease, some truss-like structural branches appear. The delicate structures can offset the shear effect in the joints. But at the same time it reduces manufacturability and structural stiffness. At this point, the choice of constraint upper bound is very important.

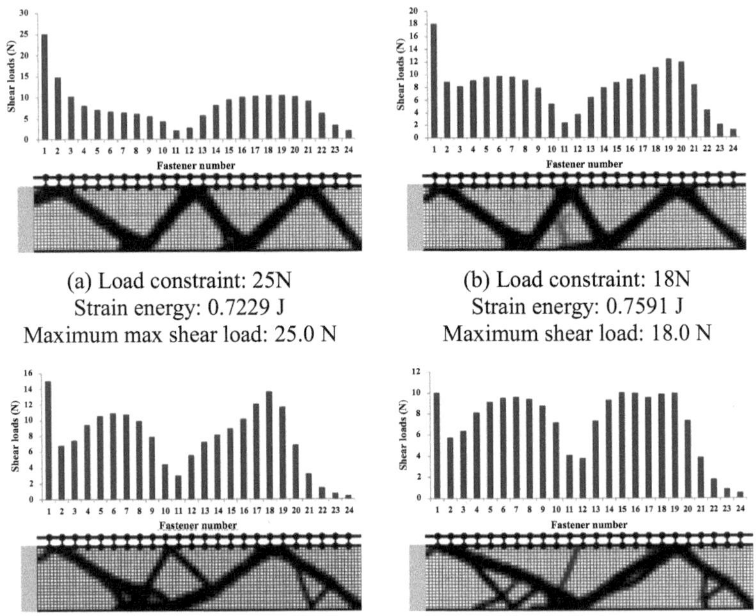


(a) Load constraint: 25N
Strain energy: 0.7229 J
Maximum max shear load: 25.0 N

(b) Load constraint: 18N
Strain energy: 0.7591 J
Maximum shear load: 18.0 N

(c) Load constraint: 15N
Strain energy: 0.7909 J
Maximum max shear load: 15.0N

(d) Load constraint: 10N
Strain energy: 0.8478 J
Maximum max shear load: 10.0N

**Figure 6.12.** *Optimized structures and load distributions with different constraints*

## 6.3.2. Two different wing boxes

Consider now a wing box structure shown in Figure 6.13. It consists of two tapered I-beams covered with two thin sheets. Both beams are clamped at one end. A distributed pressure of $5 \times 10^4$ Pa is applied on the lower skin. Due to the symmetry, only one of the two beams is considered.

Suppose skins and webs of the I-beams have a thickness of 2 mm. The beams and skins are meshed with quadrangular shell elements of size 1 mm × 1 mm and 2 mm × 2 mm, respectively. Each beam is fastened to the thin sheets with four rows of joints marked as Rows A, B, C and D. For the whole structure, the elastic modulus is $2.6 \times 10^9$ Pa and Poisson's ratio is 0.3. After the first FE analysis, the shear loads of the initial solid design are shown in Figure 6.14. We can find that the maximum shear loads occur at the root of the I-beam in all the four rows.

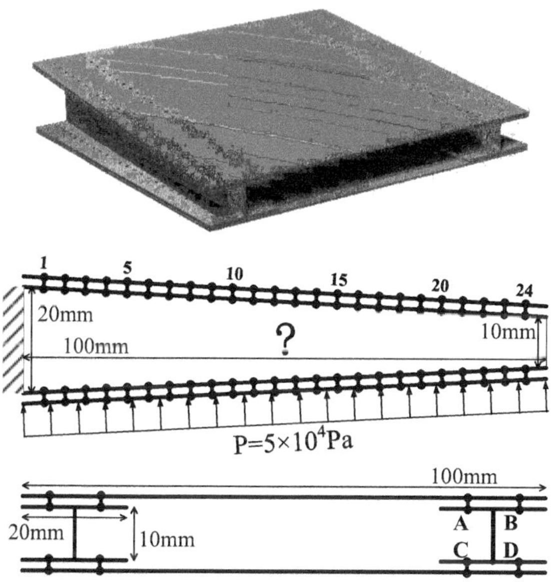


**Figure 6.13.** *Wing-box structure with four rows of joints*

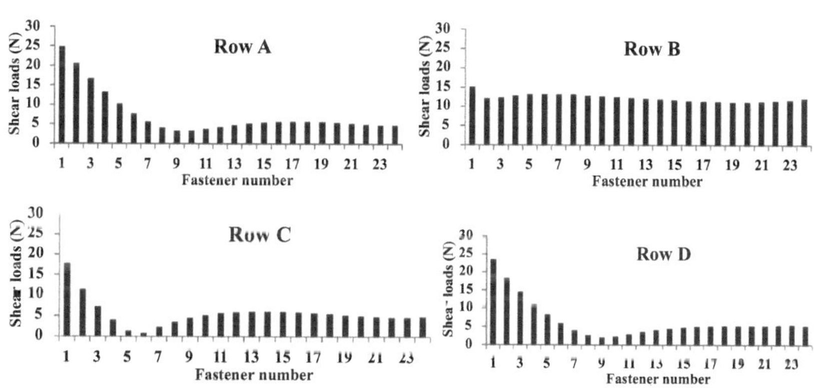


**Figure 6.14.** *Initial distribution of the joint loads in the joints*

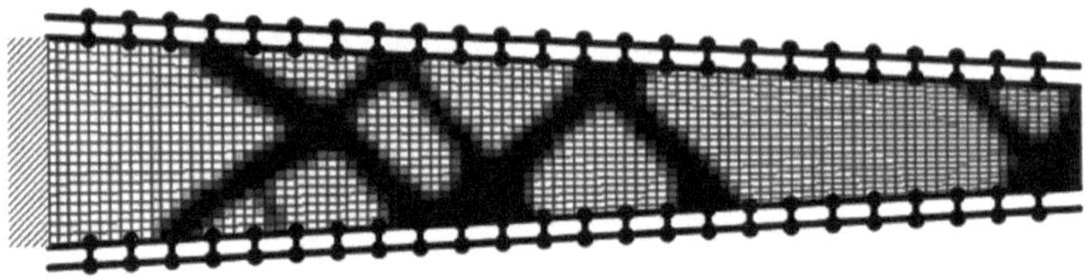

**Figure 6.15.** *The optimized design without joint load constraint (mean compliance 1.033 J, maximum joint load 27.5 N)*

Likewise, topology optimization is intended to minimize the mean compliance. The constraint of 30% volume fraction is imposed to the beam web. Similarly, two optimizations are performed for the purpose of comparison. One of them constrains the shear load in the joints to be less than 20 N. Figures 6.15 and 6.16 present the two optimized structural topologies. Corresponding joint loads are plotted and compared in Figure 6.17.

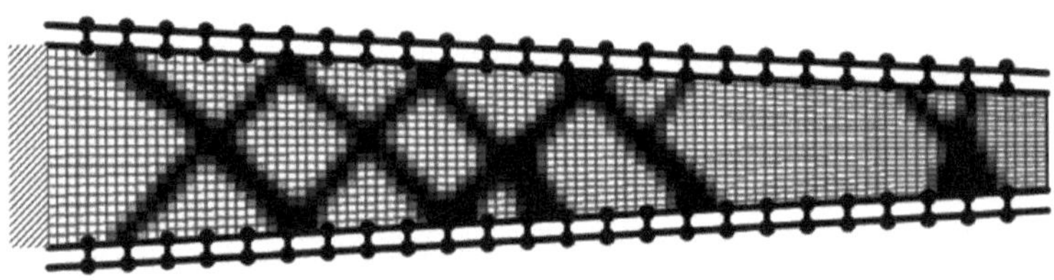

**Figure 6.16.** *The optimized design with joint load constraint (mean compliance 1.082 J, maximum joint load 20 N)*

Figures 6.15–6.17 indicate that the maximum joint load appears in row D in both cases. Compared with the initial joint loads presented in Figure 6.14, the distributions of joint loads change considerably. The comparison between Figures 6.15 and 6.16 indicates that materials are topologically distributed toward the beam root to offset the joint loads. Furthermore, the principal stresses near the beam roots in two optimized designs are plotted in Figure 6.18. This confirms that the structure is optimized to provide a push for the upper plate and a pull for the lower one near the beam root.

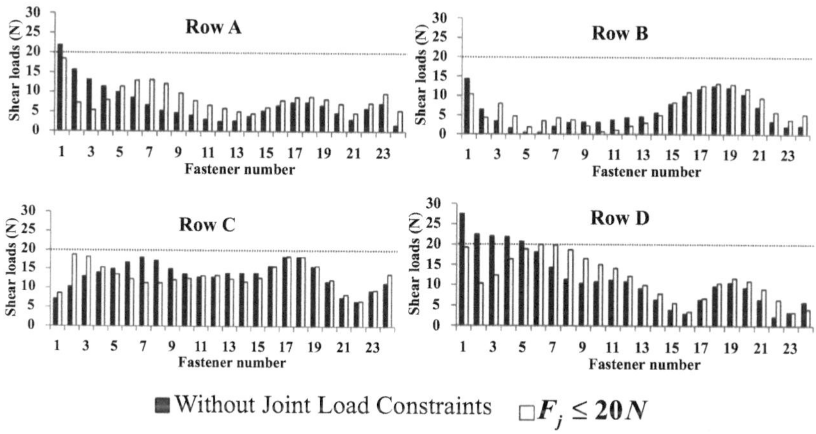


**Figure 6.17.** *Joint load distributions for two optimized designs*

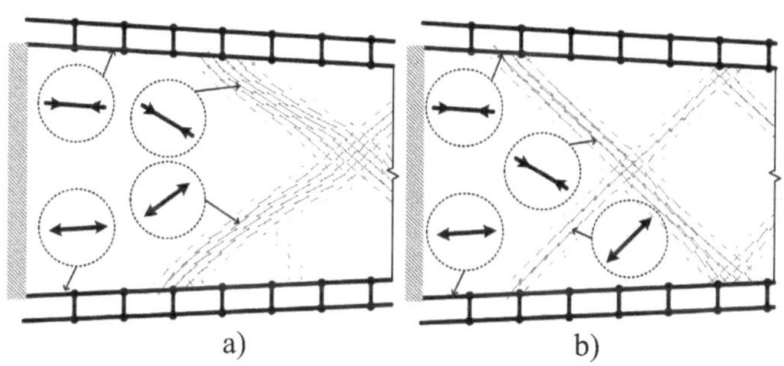


**Figure 6.18.** *Principal stresses of the optimized designs*

Finally, a flat wing of length 8.22 m and width 2.5 m as shown in Figure 6.19 is considered in the topology optimization with joint load constraints. Two typical load cases are applied, i.e. the bending and torsion conditions with the aerodynamic forces loaded on the skin of the wing. The deformations and stress distributions of the two load cases are plotted in Figure 6.20. As the maximum joint loads appear near the root of the wing, we decide to choose two wing boxes as the topological design domain, as shown in Figure 6.21.

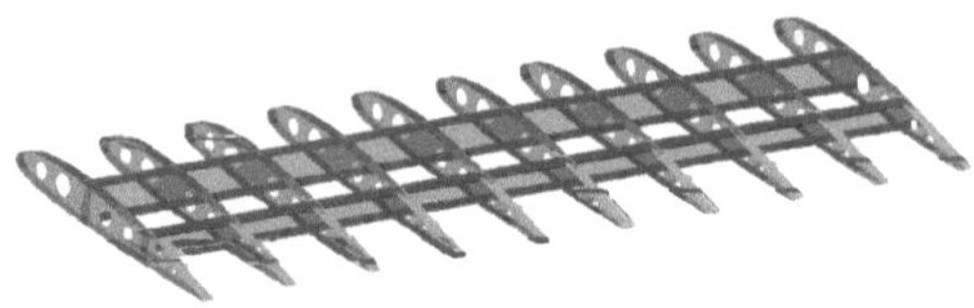

**Figure 6.19.** *Structural configuration of a flat wing*

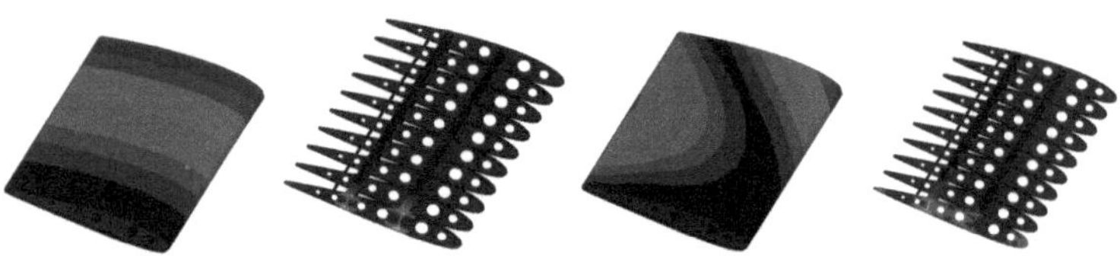

a) b)

**Figure 6.20.** *Deformations and stress distributions of a) bending and b) torsion load cases. For a color version of this figure, see www.iste.co.uk/zhang/topology.zip*

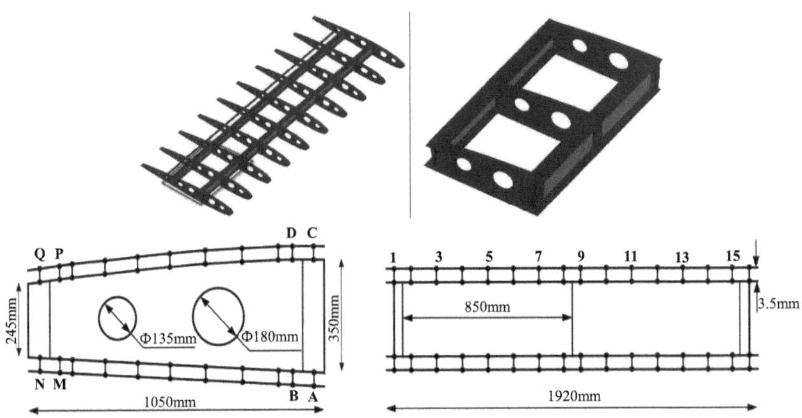


**Figure 6.21.** *Wing boxes chosen as the design domain and the configuration of the fasteners*

Analyses of the original design have shown that the joint loads in the torsion case are much larger than those of the bending case. The maximum joint load is now 17.5 KN. The distribution of the joint loads on the chosen wing box is shown in Figure 6.22.

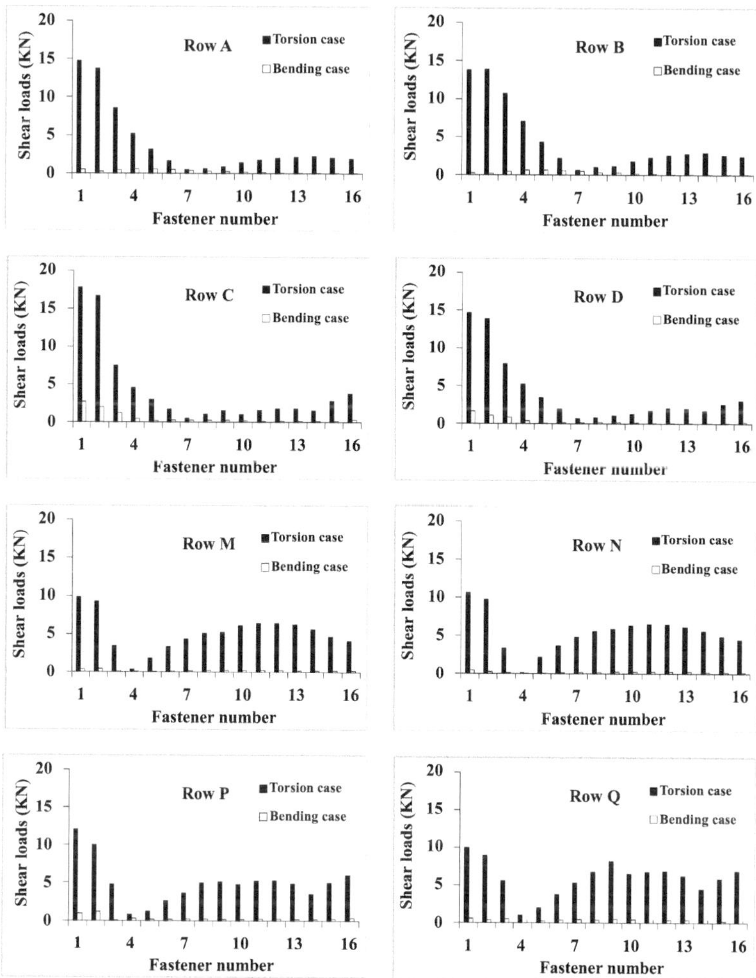


**Figure 6.22.** *Distribution of the joint loads on the chosen wing box*

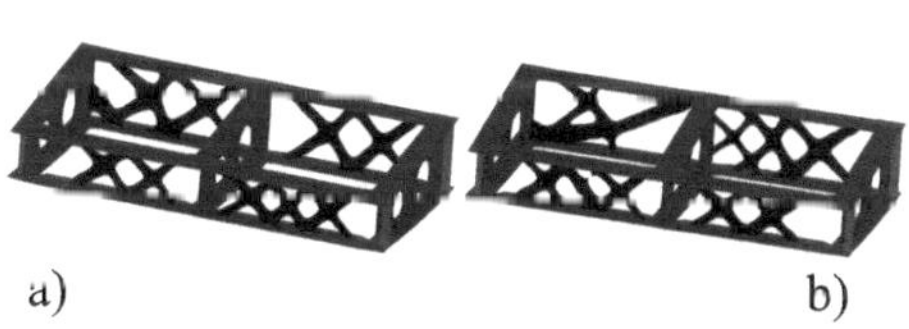


**Figure 6.23.** *Topology optimization result of: a) a standard design and; b) the design with joint load constraints*

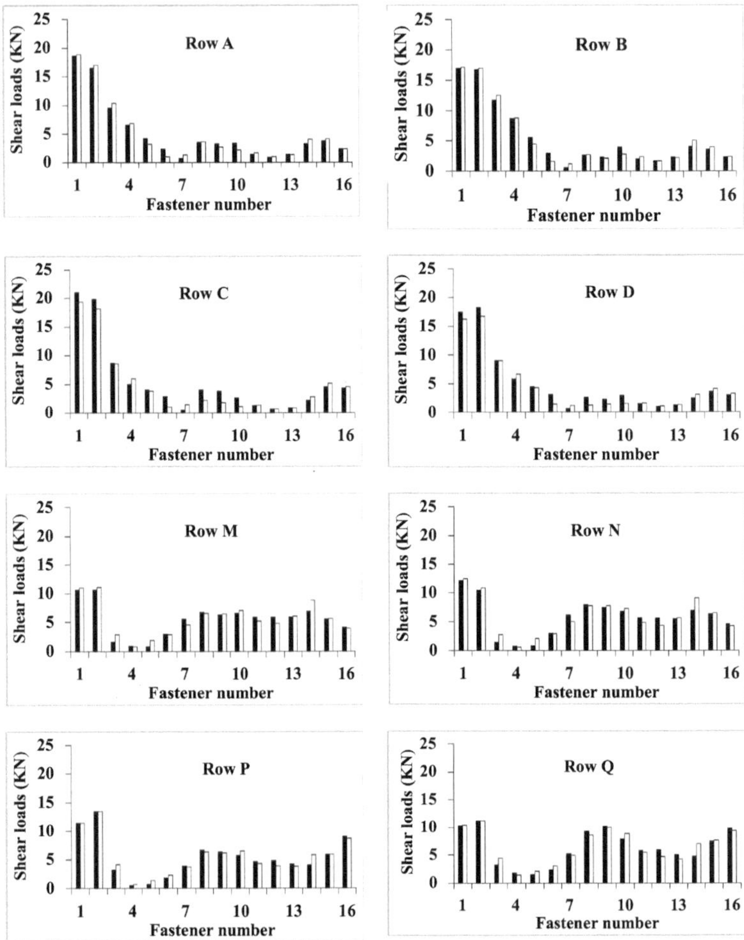


**Figure 6.24.** *Comparisons of the joint loads of the standard design (black) and those with joint load constraints (white)*

A volume constraint is imposed and bounded by a volume fraction 35% in both standard topology optimization and those with constraints on joint loads limited to be lower than 19.5 KN. The final designs are shown in Figure 6.23. Comparisons are shown in Figure 6.24 and Table 6.3.

| Items | Optimized design without joint load constraints | Optimized design with $F_j \leq 20N$ | Related difference |
|---|---|---|---|
| Mean compliance (J) | 131.9 | 133.4 | +1.1% |
| Maximum shear loads (N) | 21.01 | 19.36 | –7.9% |
| Volume fractions | 0.35 | 0.35 | 0 |

**Table 6.3.** *Comparison of optimized results*

## 6.4. Conclusions

In this chapter, we present a structural topology optimization method with constraints on the shear loads of multifastener joints, which is inspired from practical aircraft design. The joints are modeled as short beam elements. Design sensitivities of shear loads with respect to pseudo-densities are derived using the adjoint method in terms of the derivatives of nodal displacements of beam elements, where additional finite element analyses are needed.

Several numerical examples, including aircraft wing section design, are tested and optimization results are compared with standard topology optimization designs. It is shown that the joint loads are strongly affected by the structural layout and the load carrying path. With an optimized distribution of structural stiffness, all the shear loads in the joints are perfectly controlled by the prescribed upper bound.

To validate the proposed optimization method, loading tests are carried out with the help of stereolithography resin models. By comparing the minimum stretching force breaking the joints in the resin model, three groups of experiments have obviously verified the effect of the joint load constraints upon the optimized topology.

# 7

# Potential Applications of Topology Optimization

## 7.1. Shape-preserving design

In most existing work on engineering structural designs, typical conceptual designs were obtained from the best load carrying path generated by a global strain energy-based topology optimization design. Further detailed shape and sizing optimization designs were subsequently carried out to improve local performances such as strength and stability, etc. However, in many cases, it is crucial to restrain the warping deformations and maintain the coordinated displacements during the procedures of structural design, manufacturing, assembling and service [NIU 88, CAI 06, XIE 07]. The design specification is to obtain better deformation behaviors of the elastic bodies, which is more than a global strain energy design. For example, structures on the aircraft front fuselage, as shown in Figure 7.1, will be designed properly not only for strength and stiffness performance, but also to ensure a coordinate deformation of the windshield to avoid cracking. Similar design requirements can be found for the supporting structures of the large numbers of openings and components on the aircraft.

In fact, techniques of topology optimization have long been used to obtain required structural deformation patterns, which mainly results in the design of compliant mechanisms. In the existing literature, constraints on a single nodal displacement or multiple displacements were normally issued. The magnitudes of different nodal displacements were controlled to form a better deformation.

**Figure 7.1.** *Structure layout and windshield of front fuselage*

However, obtaining a coordinated displacement with least warping deformation is rather complicated and computationally expensive, with large numbers of constraints on magnitudes of nodal displacements. The key difficulty lies in how to distinguish the rigid body motion and warping deformation from the total deformation patterns.

Recently, we have proposed a shape-preserving topology optimization method. Local strain energies on specified shape preserving were assigned as design constraints. The structural total deformation consists of the warping deformation and rigid body motion. As the strain energy of the rigid body motion is completely zero, the warping deformation can be easily distinguished and suppressed in this way, as shown in Figure 7.2.

Typical designs can be found as shown in Figure 7.3. Structural configuration of a front fuselage was designed to obtain a minimum global strain energy with control of local warping deformation on the windshield in Figure 7.3(b). Compared with the standard topology optimization result shown in Figure 7.3(a), the shape-preserving design reduced the local strain energy in the windshield up to 20%, while the cost of the global strain energy increased only by 5%.

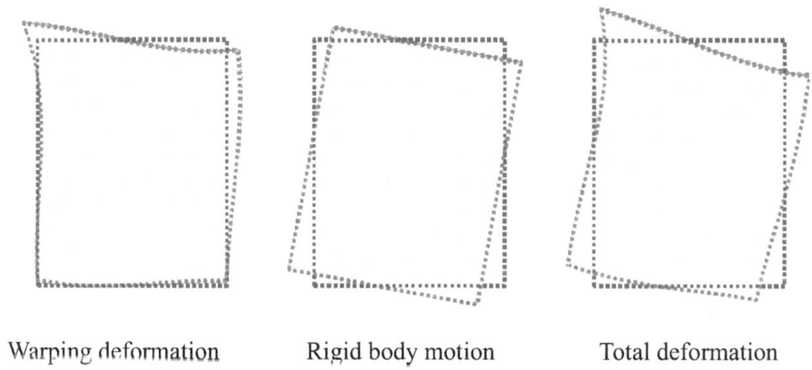


**Figure 7.2.** *Different deformation patterns*

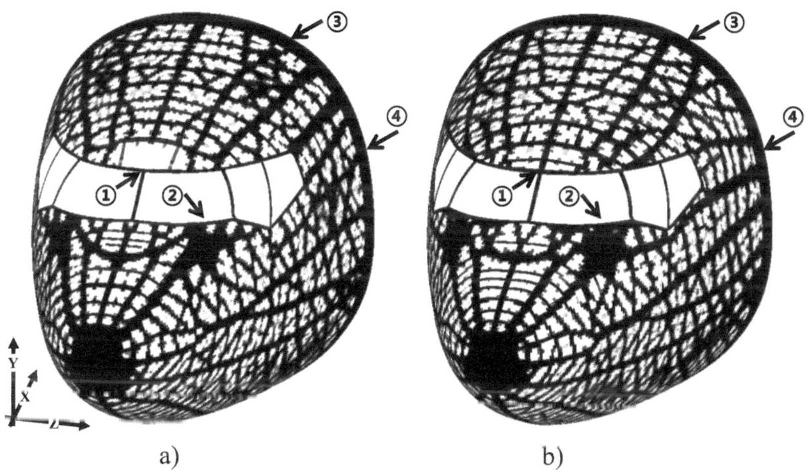


**Figure 7.3.** *Topology optimization of a front fuselage: a) standard topology optimization and b) shape-preserving design*

Potential applications of shape-preserving design can be extended to aircraft assembly jigs design and precision design of some other functional surfaces. As shown in Figure 7.4, during the structure design of assembly jigs, the shape of the supported aircraft components, especially the assembling interface, will be preserved to ensure a precise assembly. Tiny elastic warping deformations under external loading conditions and gravity must be suppressed under shape-preserving constraints. In Figure 7.5, a helicopter aiming system is designed to maintain a parallel direction for subsystems of TV, infrared and laser aiming. The thermal-elastic warping deformation will lead to non-straight aiming. The recent solution of shape-preserving topology optimization for the aiming system pedestal is also shown in Figure 7.5.

**Figure 7.4.** *A typical assembly jig for aircraft wing*

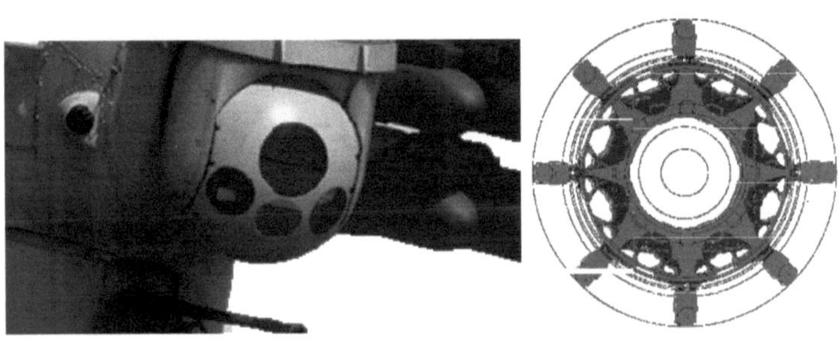

**Figure 7.5.** *A helicopter aiming system and shape-preserving design of its pedestal*

## 7.2. Smart structure design

Recently, smart structures have been recognized as one of the most important structural styles for next-generation aircrafts and aerospace vehicles. Existing attempts to use topology optimization in designing smart structures were based on aerodynamic performance, i.e. the morphing aircraft structure design. For example, Reich *et al.* [REI 07] proposed a two-step topology optimization design for morphing vehicle skins. The substructures of the aircraft skin were first considered as truss-like mechanisms. Driven by actuators, the global structure will deform to a specific goal shape. The material properties of the skin were optimized to meet the global deformation requirement. In the second step, multi-phase material microstructures were designed using topology optimization to satisfy prespecified requirements for the skin design. The effective properties of the microstructures were further evaluated with loading test.

Inoyama *et al.* [INO 08] later presented new topology optimization approaches that determine the distribution of structural properties and actuators, to obtain a morphing wing with multiple target shapes. By acquiring the constraints on the truss volume and actuators' distribution, the topological design results satisfied the design requirements effectively. Figure 7.6 shows the typical designs where three different configurations of the morphing wing for different flight cases are presented. The topological design includes the distribution of trusses, different actuators, structural components and linkages.

In topology optimization of smart structures, actuators are the key elements to realize structural morphing. These actuators may be replaced with some smart structural components, e.g. shape memory alloys (SMAs), whose solid-to-solid phase transformations induced by appropriate temperature and/or stress changes can recover structural deformations. These components have been widely used in many practical areas due to their particular properties. Early works of SMA applications have been summarized by Van Humbeeck [VAN 99]. In the work of Beauchamps *et al.* [BEA 92], SMA was used to control the surfaces of the wings and rotor blades in adaptation to different flight speed and attack angle. As a result, flight efficiency was improved and noise was reduced.

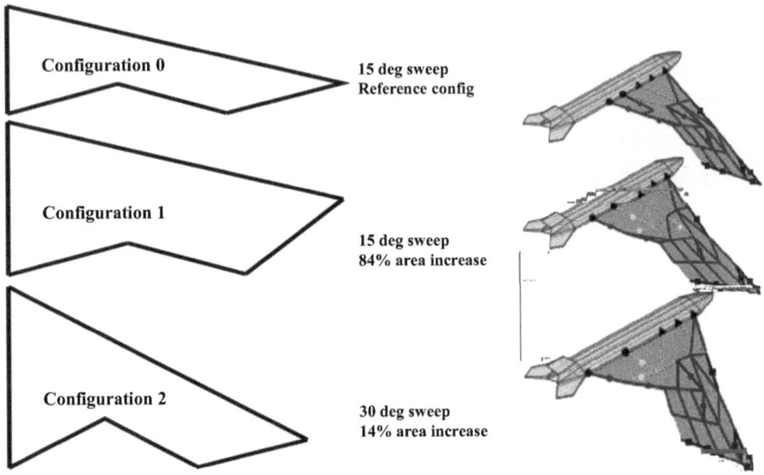


**Figure 7.6.** *Topology optimization of morphing wing [INO 08]*

Later, Kudva [KUD 04] reviewed the DARPA project on smart wing design. As shown in Figure 7.7, the leading edge of the unmanned aerial vehicle (UAV) was actuated by SMA to improve the flight efficiency. More recent applications of SMA in aircraft and aerospace designs can be found in the works of Hartl *et al.* [HAR 07, HAR 10].

As presented by Van Humbeeck [VAN 99], SMA also has more particular properties, such as high damping capacity and pseudoelastic, besides the shape memory effects. These properties bring more possibilities to future smart structure systems design, especially in topology optimization design. For example, SMA actuators or structures can first be considered as some functional components integrated with the structures to have better mechanical and mechanism properties. The integrated system can be realized using the integrated layout and topology optimization to improve the structural efficiency of deformation. Second, the configuration of SMA structures themselves can also be optimized to improve the deformation and actuation under thermal or stress inputs, such as the work done by Langelaar *et al.* [LAN 11]. Third, due to the damping properties, SMA structures can be good alternatives to current damping materials and structures. The material distribution of SMA can be optimized to improve the global dynamic responses.

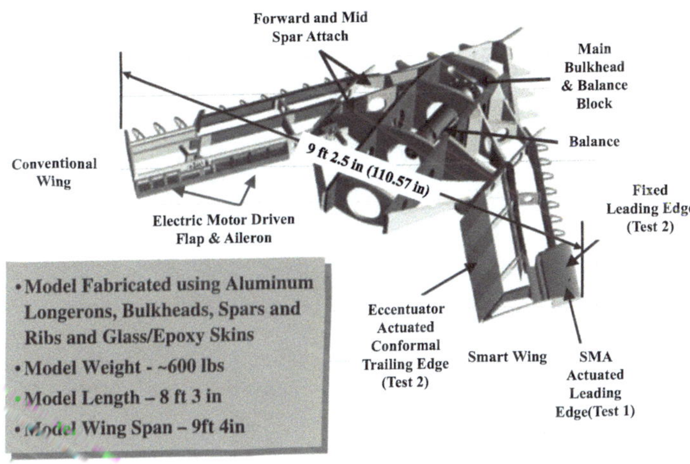


**Figure 7.7.** *Structural layout of UAV using SMA actuators [KUD 04]*

## 7.3. Structural features design

Existing topology optimization techniques, such as density-based methods and level set methods, have been developed rapidly to improve mechanical performances and obtain reasonable load-carrying paths. But to gain wide acceptance from the engineering side, more structural engineering features will be imposed in topology optimization.

Engineering features are not new in topology optimization. The abovementioned topological design with stiffeners, structural components and fasteners based on pseudo-density variables can be considered as different kinds of features design. Some other feature-based design has focused on the sizing control for level set methods topology optimization, such as the work of Mei *et al.* [MEI 08], Chen *et al.* [CHE 08] and Guo *et al.* [GUO 14a].

However, compared with the real engineering features such as stiffeners, lugs, holes and protrusions, etc., as shown in Figure 7.8, the abovementioned features are far from the requirement of practical applications. As a result, detailed designs produced by adding engineering features are needed as postprocessing procedures for existing topology optimization applications. These features can be designed for mechanical, functional or even manufacturing purposes. This is a serial design mode, which significantly changes the structural configuration and may lower the optimized structural performances.

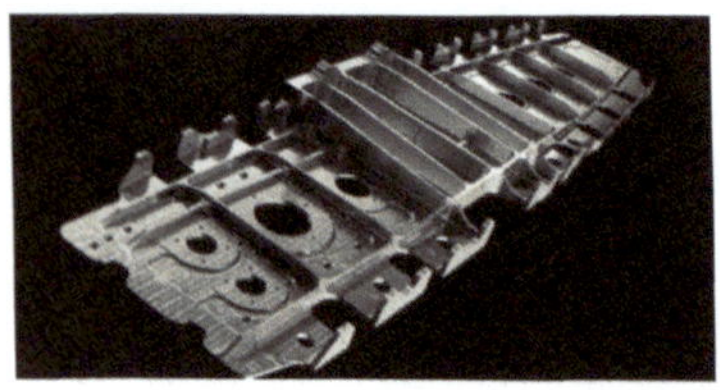

**Figure 7.8.** *Aircraft structures with engineering features. For a color version of this figure, see www.iste.co.uk/zhang/topology.zip*

Thus, in the future of topology optimization, these engineering features and their assembling patterns may need to be designed simultaneously with the material distribution, which can be considered as an integrated design for engineering features' layout and structural topology [ZHU 09, ZHU 10a]. More considerations on structural detailed performances, functional and manufacturing purposes, assembling processes, etc., will be introduced into the topological design. The optimization design of the features can therefore be achieved by improved integrated layout and topology optimization methods.

With this idea in mind, some recently proposed topology optimization methods [GUO 14b, ZHO 16] have considered features with different geometries as numbers of components. These components are described with implicit level set functions and will be joined together using Boolean operation, based on a scheme such as R-functions or KS functions. The structural topology can be obtained by the layout design of these components. A typical example is presented in Figure 7.9.

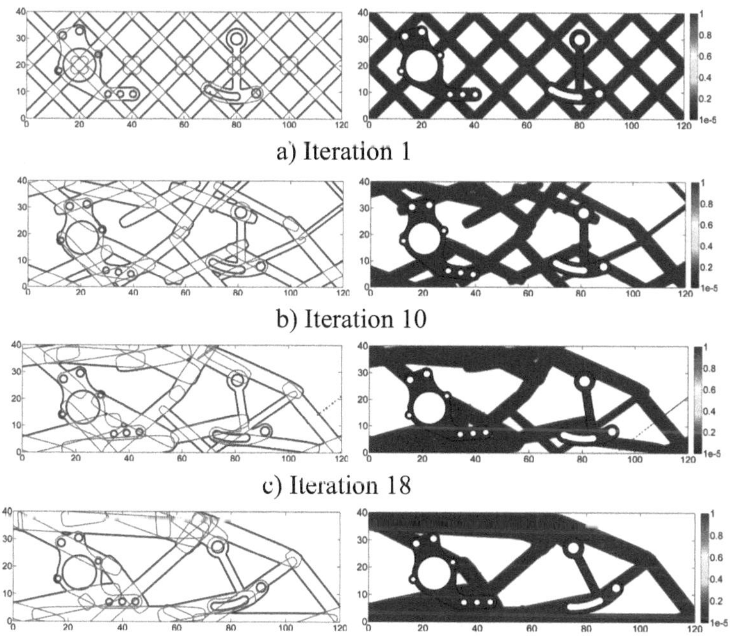

a) Iteration 1

b) Iteration 10

c) Iteration 18

d) Optimized configuration

**Figure 7.9.** *Evolutions of structural configuration involving features/components with different geometries*

## 7.4. Topology optimization and additive manufacturing

Most topology optimization researchers have been questioned about the manufacturability of their optimized structures. Before, the performance of topology optimization was compromised with some additional manufacturing constraints, such as sizing constraints, casting directions, symmetry and repeated patterns, etc. Recently, the rapidly growing additive manufacturing techniques, also known as 3D printing, which directly fabricate structures from a CAD model, change the situation and may prove to be beneficial for both sides. On the one hand, additive manufacturing techniques currently provide possibly the best manufacturing solutions for topology optimization. On the other hand, additive manufacturing needs something to prove its abilities of forming very complicated structures. Topology optimization is among the best choices.

The combination of topology optimization and additive manufacturing immediately gained preliminary success. As shown in Figure 7.10, researchers from EADS attempted to optimize the aircraft structures to have a light-weight and stiffness design. The final design was revealed with metallic additive manufacturing. The global procedure saved up to 64%, in weight with all the mechanical performances satisfied.

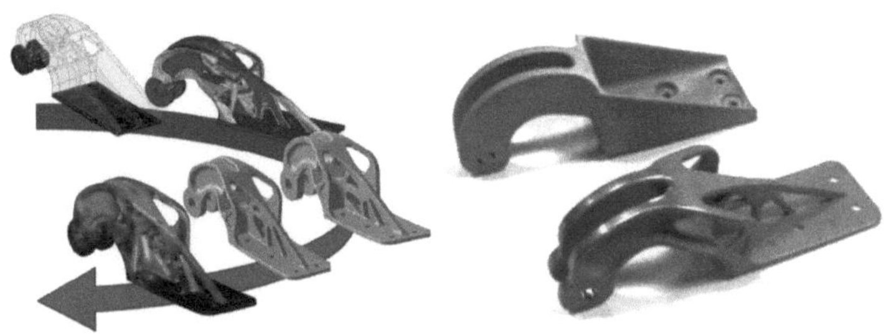

**Figure 7.10.** *Topology optimization and additive manufacturing of Airbus A320 nacelle hinge bracket [TOM 11]*

**Figure 7.11.** *Loading test with stereolithography resin model of additive manufacturing*

Metallic additive manufacturing is powerful in forming practical structural parts, but it suffers from the high cost of the equipment and processing procedure. Very recently, resinous additive manufacturing, i.e. the high efficient stereolithography rapid prototyping, is being used to validate the topology optimization results, as shown in Figures 7.11 and 7.12. It was found that the rapid prototypes of resin material can approximately maintain linear elastic and isotropic mechanical properties in a specific deformation range. Moreover, compared with metallic additive manufacturing, the resin model is much more easily and effectively fabricated.

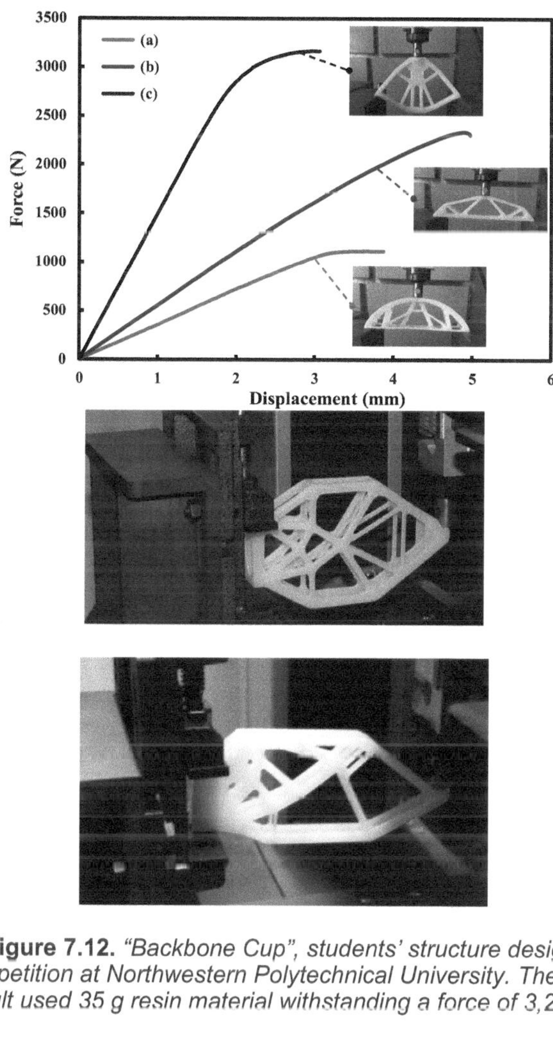


**Figure 7.12.** *"Backbone Cup", students' structure design competition at Northwestern Polytechnical University. The best result used 35 g resin material withstanding a force of 3,235 N*

In the upcoming future, it is believed that aircraft and aerospace structures, especially most UAV structures, will be designed and fabricated as unconventional integral structures to save weight and simplify the assembling procedure. With this new concept, the combination of topology optimization and additive manufacturing will surely play an important role in developing high-performance and lightweight structure systems.

# Bibliography

[AIN 01] AINSWORTH M., "Essential boundary conditions and multi-point constraints in finite element analysis", *Computer Methods in Applied Mechanics and Engineering*, vol. 190, pp. 6323–6339, 2001.

[ALL 04a] ALLAIRE G., JOUVE F., MAILLOT H., "Topology optimization for minimum stress design with the homogenization method", *Structural and Multidisciplinary Optimization*, vol. 28, pp. 87–98, 2004.

[ALL 04b] ALLAIRE G., JOUVE F., TOADER A.M., "Structural optimization using sensitivity analysis and a level-set method", *Journal of Computational Physics*, vol. 194, pp. 363–393, 2004.

[ALV 97] ALVIN K.F., "Efficient computation of eigenvector sensitivities for structural dynamics", *AIAA Journal*, vol. 35, pp. 1760–1766, 1997.

[BAR 92] BARRETT R.T., Fastener Design Manual, NASA Reference Publication, pp. 12–28, 1992.

[BAZ 06] BAZILEVS Y., BEIRAO DA VEIGA L., COTTRELL J.A. *et al.*, "Isogeometric analysis: approximation, stability and error estimates for h-refined meshes", *Mathematical Models and Methods in Applied Sciences*, vol. 16, pp. 1031–1090, 2006.

[BEA 92] BEAUCHAMPS C.H., NADOLINK R.H., DICKINSON S.C. *et al.*, "Shape memory alloy adjustable camber (SMAAC) control surfaces", *Proceedings 1st European Conference on Smart Structures and Materials*, SPIE, Glasgow, vol. 189, pp. 17–77, 1992.

[BEC 97] BECKERS M., Optimisation de structures en variables discrètes, PhD Thesis, University of Liège, 1997.

[BEC 99] BECKERS M., "Topology optimization using a dual method with discrete variables", *Structural and Multidisciplinary Optimization*, vol. 17, pp. 14–24, 1999.

[BEN 88] BENDSØE M.P., KIKUCHI N., "Generating optimal topologies in structural design using a homogenization method", *Computer Methods in Applied Mechanics and Engineering*, vol. 71, pp. 197–224, 1988

[BEN 89] BENDSØE M.P., "Optimal shape design as a material distribution problem", *Structural and Multidisciplinary Optimization*, vol. 10, pp. 193–202, 1989.

[BEN 90] BENDSØE M.P., TRIANTAFYLLIDIS N., "Scale effects in the optimal design of a microstructured medium against buckling", *International Journal of Solids and Structures*, vol. 26, pp. 725–741, 1990.

[BEN 95] BENDSØE M.P., *Optimization of Structural Topology, Shape and Material*, Springer, Berlin, Heidelberg, New York, 1995.

[BEN 99] BENDSØE M.P., SIGMUND O., "Material interpolation schemes in topology optimization", *Archive of Applied Mechanics*, vol. 69, pp. 635–654, 1999.

[BEN 02] BENDSØE M.P., "Recent developments in topology design of materials and mechanisms", *ESAIM: Proceedings*, vol. 11, pp. 41–60, 2002.

[BEN 03] BENDSØE M.P., SIGMUND O., *Topology Optimization: Theory, Method and Application*, Springer, Berlin, Heidelberg, New York, 2003.

[BES 13] BESSELINK B. TABAK U., LUTOWSKA A. *et al.*, "A comparison of model reduction techniques from structural dynamics, numerical mathematics and systems and control", *Journal of Sound and Vibration*, vol. 332, pp. 4403–4422, 2013.

[BIA 07] BIANCHI G., AGLIETTI G.S., RICHARDSON G., "Optimization of bolted joints connecting honeycomb panels", *1st CEAS, 10th European Conference on Spacecraft Structures, Materials and Mechanical Testing*, Berlin, Germany, pp. 10–13, September 2007.

[BOU 01] BOURDIN B., "Filters in topology optimization", *International Journal for Numerical Methods in Engineering*, vol. 50, pp. 2143–2158, 2001.

[BRU 01] BRUNS T.E., TORTORELLI D.A., "Topology optimization of non-linear elastic structures and compliant mechanisms", *Computer Methods in Applied Mechanics and Engineering*, vol. 190, pp. 3443–3459, 2001.

[BRU 05] BRUYNEEL M., DUYSINX P., "Note on topology optimization of continuum structures including self-weight", *Structural and Multidisciplinary Optimization*, vol. 29, pp. 245–256, 2005.

[BRU 11] BRUYNEEL M., "SFP – a new parameterization based on shape functions for optimal material selection: application to conventional composite plies", *Structural and Multidisciplinary Optimization*, vol. 43, pp. 17–27, 2011.

[BUC 98] BUCHER I., "Parametric Optimization of Structures Under Combined Base Motion Direct Forces and Static Loading", *Journal of Vibration and Acoustics*, vol. 124, no. 1, pp. 132–140, 1998.

[BUH 01] BUHL T., "Simultaneous topology optimization of structure and supports", *Structural and Multidisciplinary Optimization*, vol. 23, pp. 336–346, 2001.

[BUR 05] BURGUEÑO R., QUAGLIATA M.J., MOHANTY A.K. *et al.*, "Hierarchical cellular designs for load-bearing biocomposite beams and plates", *Materials Science and Engineering A*, vol. 390, pp. 178–187, 2005.

[CAG 98] CAGAN J., DEGENTESH D., YIN S., "A simulated annealing-based algorithm using hierarchical models for general three-dimensional component layout", *Computer-Aided Design*, vol. 30, pp. 781–790, 1998.

[CAG 02] CAGAN J., SHIMADA K., YIN S., "A survey of computational approaches to three-dimensional layout optimization", *Computer-Aided Design*, vol. 34, pp. 597–611, 2002.

[CAI 06] CAI W.W., HSIEH C.C., LONG Y.F. *et al.*, "Digital panel assembly methodologies and applications for compliant sheet components", *Journal of Manufacturing Science and Engineering – Transactions of the ASME*, vol. 128, pp. 270–279, 2006.

[CAM 91] CAMERON S., "Approximation hierarchies and s-bounds", *Proceedings Symposium on Solid Modeling Foundations and CAD/CAM Applications*, pp. 129–137, 1991.

[CAM 97] CAMANHO P.P., MATTHEWS F.L., "Stress analysis and strength prediction of mechanically fastened joints in FRP: a review", *Composites Part A: Applied Science and Manufacturing*, vol. 28, pp. 529–547, 1997.

[CHE 81] CHENG G.D., OLHOFF N., "An investigation concerning optimal design of solid elastic plates", *International Journal of Solids and Structures*, vol. 17, pp. 305–323, 1981.

[CHE 98] CHEN M.-T., ALI A., "An efficient and robust integration technique for applied random vibration analysis", *Computers and Structures*, vol. 66, pp. 785–798, 1998.

[CHE 01] CHEN B.C., KIKUCHI N., "Topology optimization with design-dependent loads", *Finite Elements in Analysis and Design*, vol. 37, pp. 57–70, 2001.

[CHE 07] CHEN J., SHAPIRO V., SURESH K. *et al.*, "Shape optimization with topological changes and parametric control", *International Journal for Numerical Methods in Engineering*, vol. 71, pp. 313–346, 2007.

[CHE 08] CHEN S., WANG M.Y., LIU A.Q., "Shape feature control in structural topology optimization", *Computer-Aided Design*, vol. 40, pp. 951–962, 2008.

[CHI 97] CHICKERMANE H., GEA H.C., "Design of multi-component structural systems for optimal layout topology and joint locations", *Engineering with Computers*, vol. 13, pp. 235–243, 1997.

[CHI 99] CHICKERMANE H., GEA H.C., YANG R.J. *et al.*, "Optimal fastener pattern design considering bearing loads", *Structural Optimization*, vol. 17, pp. 140–146, 1999.

[CHI 10] CHINTAPALLI S., ELSAYED M.S.A., SEDAGHATI R. *et al.*, "The development of a preliminary structural design optimization method of an aircraft wing-box skin-stringer panels", *Aerospace Science and Technology*, vol. 14, no. 3, pp. 188–198, 2010.

[CHO 05] CHO S., CHOI J.-Y., "Efficient topology optimization of thermo-elasticity problems using coupled field adjoint sensitivity analysis method", *Finite Elements Analysis and Design*, vol. 41, pp. 1481–1495, 2005.

[CLO 75] CLOUGH R.W., PENZIEN J., *Dynamics of Structures*, McGraw-Hill, New York, 1975.

[COE 08] COELHO P.G., FERNANDES P.R., GUEDES J.M. *et al.*, "A hierarchical model for concurrent material and topology optimization of three-dimensional structures", *Structural and Multidisciplinary Optimization*, vol. 35, pp. 107–115, 2008.

[COR 83] CORNWELL R.E., CRAIG R.R., JOHNSON C.P., "On the application of the mode-acceleration method to structural engineering problems", *Earthquake Engineering and Structural Dynamics*, vol. 11, pp. 679–688, 1983.

[DAI 08] DAI G.M., ZHANG W.H., "Size effects of basic cell in static analysis of sandwich beams", *International Journal of Solids and Structures*, vol. 45, pp. 2512–2533, 2008.

[DEA 13] DEATON J.D., GRANDHI R.V., "Stiffening of restrained thermal structures via topology optimization", *Structural and Multidisciplinary Optimization*, vol. 48, pp. 731–745, 2013.

[DEA 14] DEATON J.D., GRANDHI R.V., "A survey of structural and multidisciplinary continuum topology optimization, post 2000", *Structural and Multidisciplinary Optimization*, vol. 49, pp. 1–38, 2014.

[DEB 88] DE BONT F.M.J., AARTS E.H.L., MEEHAN P. *et al.*, "Placement of shapeable blocks", *Philips Journal of Research*, vol. 43, pp. 1–22, 1988.

[DEN 13] DENG J., YAN J., CHENG G., "Multi-objective concurrent topology optimization of thermoelastic structures composed of homogeneous porous material", *Structural and Multidisciplinary Optimization*, vol. 47, pp. 583–597, 2013.

[DER 04] DE RUITER M.J., VAN KEULEN F., "Topology optimization using a topology description function", *Structural and Multidisciplinary Optimization*, vol. 26, pp. 1615–1488, 2004.

[DES 13] DESMORAT B., "Structural rigidity optimization with an initial design dependent stress field. Application to thermo-elastic stress loads", *European Journal of Mechanics – A/Solid*, vol. 37, pp. 150–159, 2013.

[DÍA 92] DÍAZ A.R., KIKUCHI N., "Solutions to shape and topology eigenvalue optimization problems using a homogenization method", *International Journal for Numerical Methods in Engineering*, vol. 35, pp. 1487–1502, 1992.

[DIC 97] DICKENS J.M., NAKAGAWA J.M., WITTBRODT M.J., "A critique of mode acceleration and modal truncation augmentation methods for modal response analysis", *Computers and Structures*, vol. 62, pp. 985–998, 1997.

[DU 07] DU J., OLHOFF N., "Topological design of freely vibrating continuum structures for maximum values of simple and multiple eigenfrequencies and frequency gaps", *Structural and Multidisciplinary Optimization*, vol. 34, pp. 91–110, 2007.

[EDW 07] EDWARDS C.S., KIM H.A., BUDD C.J., "An evaluative study on ESO and SIMP for optimizing a cantilever tie-beam", *Structural and Multidisciplinary Optimization*, vol. 34, pp. 403–414, 2007.

[EKH 08] EKH J., SCHÖN J., "Finite element modeling and optimization of load transfer in multi-fastener joints using structural elements", *Composite Structures*, vol. 82, no. 2, pp. 245–256, 2008.

[ESC 94] ESCHENAUER H.A., KOBELEV H.A., SCHUMACHER A., "Bubble method for topology and shape optimization of structures", *Structural Optimization*, vol. 8, pp. 42–51, 1994.

[ESC 01] ESCHENAUER H.A., OLHOFF N., "Topology optimization of continuum structures: a review", *Applied Mechanics Reviews*, vol. 54, pp. 331–390, 2001.

[FLE 86] FLEURY C., BRAIBANT V., "Structural optimization: a new dual method using mixed variables", *International Journal for Numerical Methods in Engineering*, vol. 23, pp. 409–428, 1986.

[FLE 89] FLEURY C., "First and second order convex approximation strategies in structural optimization", *Structural Optimization*, vol. 1, pp. 3–10, 1989.

[FUJ 01] FUJII D., CHEN B., KIKUCHI N., "Composite material design of two-dimensional structures using the homogenization design method", *International Journal for Numerical Methods in Engineering*, vol. 50, pp. 2031–2051, 2001.

[GAO 08] GAO T., ZHANG W., ZHU J.H., "Topology optimization of heat conduction problem involving design-dependent heat load effect", *Finite Elements in Analysis and Design*, vol. 44, no. 14, pp. 805–813, 2008.

[GAO 10] GAO T., ZHANG W., "Topology optimization involving thermo-elastic stress loads", *Structural and Multidisciplinary Optimization*, vol. 42, pp. 725–738, 2010.

[GAO 11] GAO T., ZHANG W., "A mass constraint formulation for structural topology optimization with multiphase materials", *International Journal for Numerical Methods in Engineering*, vol. 88, pp. 774–796, 2011.

[GAO 12] GAO T., ZHANG W., DUYSINX P., "A bi–value coding parameterization scheme for the discrete optimal orientation design of the composite laminate", *International Journal for Numerical Methods in Engineering*, vol. 91, pp. 98–114, 2012.

[GER 05] GERSBORG-HANSEN A., BENDSØE M.P., HABER R.B., "Topology optimization of Channel flow problems", *Structural and Multidisciplinary Optimization*, vol. 30, pp 181–192, 2005.

[GER 06] GERSBORG-HANSEN A., BENDSØE M.P., SIGMUND O., "Topology optimization of heat conduction problems using the finite volume method", *Structural and Multidisciplinary Optimization*, vol. 31, pp. 251–259, 2006.

[GRI 94] GRIMES R.G., LEWIS J.G., SIMON H.D., "A shifted block Lanczos algorithm for solving sparse symmetric generalized eigenproblems", *SIAM Journal on Matrix Analysis and Applications*, vol. 15, pp. 228–272, 1994.

[GU 00] GU J.M., MA Z.D., HULBERT M.G., "A new load-dependent Ritz vector method for structural dynamics analyses, quasi-static Ritz vectors", *Finite Elements in Analysis and Design*, vol. 36, pp. 261–278, 2000.

[GU 12] GU X.J., ZHU J.H., ZHANG W., "The lattice structure configuration design for stereolithography investment casting pattern using topology optimization", *Rapid Prototyping Journal*, vol. 18, no. 5, pp. 353–361, 2012.

[GUE 90] GUEDES J.M., KIKUCHI N., "Preprocessing and postprocessing for materials based on the homogenization method with adaptive finite element methods", *Computer Methods in Applied Mechanics and Engineering*, vol. 83, pp. 143–198, 1990.

[GUE 04] GUEST J.K., PRÉVOST J.H., BELYTSCHKO T., "Achieving minimum length scale in topology optimization using nodal design variables and projection functions", *International Journal for Numerical Methods in Engineering*, vol. 61, no. 2, pp. 238–254, 2004.

[GUO 10] GUO X., CHENG G.-D., "Recent development in structural design and optimization", *Acta Mechanica Sinica*, vol. 26, pp. 807–823, 2010

[GUO 14a] GUO X., ZHANG W., ZHONG W., "Explicit feature control in structural topology optimization via level set method", *Computer Methods in Applied Mechanics and Engineering*, vol. 272, pp. 354–378, 2014.

[GUO 14b] GUO X., ZHANG W., ZHONG W., "Topology optimization based on moving deformable components: a new computational framework", *ASME Journal of Applied Mechanics*, vol. 81, no. 8, pp. 081009–081012, 2014.

[HAB 96] HABER R.B., JOG C.S., BENDSØE M.P., "A new approach to variable-topology shape design using a constraint on perimeter", *Structural and Multidisciplinary Optimization*, vol. 11, pp. 1–12, 1996.

[HAR 07] HARTL D.J., LAGOUDAS D.C., "Aerospace applications of shape memory alloys", *Journal of Aerospace Engineering*, vol. 221, pp. 535–552, 2007.

[HAR 10] HARTL D.J., LAGOUDAS D.C., CALKINS F.T. *et al.*, "Use of a Ni60Ti shape memory alloy for active jet engine chevron application: I. Thermomechanical characterization", *Smart Materials and Structures*, vol. 19, pp. 427–437, 2010.

[HOR 99] HORI M., NEMAT-NASSER S., "On two micromechanics theories for determining micro-macro relation in heterogeneous solids", *Mechanics of Materials*, vol. 31, pp. 667–682, 1999.

[HUA 07] HUANG W.Q., CHEN D.B., XU R.C., "A new heuristic algorithm for rectangle packing", *Computers and Operations Research*, vol. 34, pp. 3270–3280, 2007.

[HUA 08] HUANG X., XIE Y.M., "A new look at ESO and BESO optimization methods", *Structural and Multidisciplinary Optimization*, vol. 35, pp. 89–92, 2008.

[HUA 09] HUANG X., XIE Y.M., "Bi-directional evolutionary topology optimization of continuum structures with one or multiple materials", *Computational Mechanics*, vol. 43, pp. 393–401, 2009.

[HUB 93] HUBBARD P.M., "Interactive collision detection", *Proceedings of IEEE Symposium on Research Frontiers in Virtual Reality*, pp. 24–32, 1993.

[INO 08] INOYAMA D., SANDERS B.P., JOO J.J., "Topology optimization approach for the determination of the multiple-configuration morphing wing structure", *Journal of Aircraft*, vol. 45, pp. 1853–1862, 2008.

[JEN 07] JENSEN J.S., "Topology optimization of dynamics problems with Padé approximants", *International Journal for Numerical Methods in Engineering*, vol. 72, pp. 1605–1630, 2007.

[JIA 97] JIANG T., CHIREHDAST M., "A system approach to structural topology optimization: designing optimal connections", *Journal of Mechanical Design*, vol. 119, pp. 40–47, 1997.

[JOG 96] JOG C.S., HABER R.B., "Stability of finite element models for distributed-parameter optimization and topology design", *Computer Methods in Applied Mechanics and Engineering*, vol. 130, pp. 203–226, 2002.

[JOG 02] JOG C.S., "Topology design of structures subjected to periodic loading", *Journal of Sound and Vibration*, vol. 253, pp. 687–709, 2002.

[JUN 06] JUNG D., GEA H., "Design of an energy-absorbing structure using topology optimization with a multimaterial model", *Structural and Multidisciplinary Optimization*, vol. 32, pp. 251–257, 2006.

[KAN 12] KANG Z., ZHANG X., JIANG S. *et al.*, "On topology optimization of damping layer in shell structures under harmonic excitations", *Structural and Multidisciplinary Optimization*, vol. 46, pp. 51–67, 2012.

[KIM 03] KIM H.A., QUERIN O.M., STEVEN G.P., "Improving efficiency of evolutionary structural optimization by implementing fixed grid mesh", *Structural and Multidisciplinary Optimization*, vol. 24, pp. 441–448, 2003.

[KRO 02] KROG L., TUCKER A., ROLLEMA G., "Application of topology, sizing and shape optimization methods to optimal design of aircraft components", *Proceedings of 3rd Altair UK HyperWorks Users Conference*, pp. 1–13, 2002.

[KRO 04] KROG L., TUCKER A., KEMP M. *et al.*, "Topology optimization of aircraft wing box ribs", *10th AIAA/ISSMO Multidisciplinary Analysis and Optimization Conference*, pp 1–11, 2004.

[KUD 04] KUDVA J.N., "Overview of the darpa smart wing project", *Journal of Intelligent Material Systems and Structures*, vol. 15, no. 4, pp. 261–267, 2004.

[LAI 08] LAI K.L., CRASSIDIS J., "Extensions of the first and second complex-step derivative approximations", *Journal of Computational and Applied Mathematics*, vol. 219, pp. 276–293, 2008.

[LAN 11] LANGELAAR M., YOON G.H., KIM Y.Y. *et al.*, "Topology optimization of planar shape memory alloy thermal actuators using element connectivity parameterization", *International Journal for Numerical Methods in Engineering*, vol. 88, pp. 817–840, 2011.

[LEM 07] LEMAIRE E., ROCHUS V., FLEURY C. *et al.*, "Topology optimization of microbeams including layer deposition manufacturing constraints", *7th World Congress on Structural and Multidisciplinary Optimization*, Seoul, pp. 1–10, 2007.

[LI 99] LI Q., STEVEN G.P., XIE Y.M., "Displacement minimization of thermoelastic structures by evolutionary thickness design", *Computer Methods in Applied Mechanics and Engineering*, vol. 179, pp. 361–378, 1999.

[LI 01] LI Q., STEVEN G.P., XIE Y.M., "Evolutionary structural optimization for connection topology design of multi-component systems", *Engineering Computations*, vol. 18, pp. 460–479, 2001.

[LIN 85] LIN J.H., "A deterministic method for the computation of stochastic earthquake response", *Earthquake Engineering and Engineering Vibration*, vol. 5, pp. 89–94, 1985.

[LIN 92] LIN J.H., "A fast CQC algorithm of psd matrices for random seismic responses", *Computers Structures*, vol. 44, pp. 683–687, 1992.

[LIN 98] LIN M., GOTTSCHALK S., "Collision detection between geometric models: a survey", *Proceeding of IMA Conference on Mathematics of Surfaces*, pp. 1–20, 1998.

[LIN 01] LIN J.H., ZHAO Y., ZHANG Y.H., "Accurate and highly efficient algorithms for structural stationary/non-stationary random responses", *Computer Methods in Applied Mechanics and Engineering*, vol. 191, pp. 103–111, 2001.

[LIN 11] LIN Z.-Q., GEA H.C., LIU S.-T., "Design of piezoelectric energy harvesting devices subjected to broadband random vibrations by applying topology optimization", *Acta Mech Sinica*, vol. 27, pp. 730–737, 2011.

[LIU 08] LIU L., YAN J., CHENG G., "Optimum structure with homogeneous optimum truss-like material", *Computers and Structures*, vol. 86, pp. 1417–1425, 2008.

[LIU 15] LIU H., ZHANG W., GAO T., "A comparative study of dynamic analysis methods for structural topology optimization under harmonic force excitations", *Structural and Multidisciplinary Optimization*, vol. 51, pp. 1321–1333, 2015.

[LU 09] LU F., LIN J.H., KENNEDY D. *et al.*, "An algorithm to study non-stationary random vibrations of vehicle–bridge systems", *Computers & Structures*, vol. 87, no. 3, pp. 177–185, 2009.

[LUO 04] LUO Z., CHENG L.P., HUANG Y.Y. *et al.*, "Topology optimization design for continuum structures", *Advances in Mechanics*, vol. 34, pp. 463–476, 2004.

[LUO 07] LUO Z., TONG L., WANG M.Y. *et al.*, "Shape and topology optimization of compliant mechanisms using a parameterization level set method", *Journal of Computational Physics*, vol. 227, pp. 680–705, 2007.

[LUO 08] LUO Z., WANG M.Y., WANG S. *et al.*, "A level set based parameterization method for structural shape and topology optimization", *International Journal for Numerical Methods in Engineering*, vol. 76, pp. 1–26, 2008.

[MA 95] MA Z.D., KIKUCHI N., CHENG H.C., "Topological design for vibrating structures", *Computer Methods in Applied Mechanics and Engineering*, vol. 121, pp. 259–280, 1995.

[MA 11] MA J., GAO W., WRIGGERS P. *et al.*, "Structural dynamic optimal design based on dynamic reliability", *Engineering Structures*, vol. 33, no. 2, pp. 468–476, 2011.

[MAR 03] MARTINS J.R., STURDZA P., ALONSO J.J., "The complex-step derivative approximation", *ACM Transactions on Mathematical Software TOMS*, vol. 29, pp. 245–262, 2003.

[MAR 05] MARTINS J., POON N.M., "On structural optimization using constraint aggregation", *VI World Congress on Structural and Multidisciplinary Optimization WCSMO6*, Rio de Janeiro, Brasil, Citeseer, pp. 1–10, 2005.

[MAU 04] MAUTE K., ALLEN M., "Conceptual design of aeroelastic structures by topology optimization", *Structural and Multidisciplinary Optimization*, vol. 27, pp. 27–42, 2004.

[MEA 82] MEAGHER D., "Geometric modeling using octree encoding", *Computer Graphics and Image Processing*, vol. 19, pp. 129–147, 1982.

[MEI 04a] MEI Y., WANG X., "A level set method for structural topology optimization with multi constraints and multi-materials", *Acta Mechanica Sinica*, vol. 20, pp. 507–518, 2004.

[MEI 04b] MEI Y., WANG X.M., "A level set method for structural topology optimization and its applications", *Advances in Engineering Software*, vol. 35, pp. 415–441, 2004.

[MEI 08] MEI Y., WANG X., CHENG G., "A feature-based topological optimization for structure design", *Advances in Engineering Software*, vol. 39, pp. 71–87, 2008.

[MIS 13] MISHRA S.K., ROY B.K., CHAKRABORTY S., "Reliability-based-design-optimization of base isolated buildings considering stochastic system parameters subjected to random earthquakes", *International Journal of Mechanical Sciences*, vol. 75, pp. 123–133, 2013.

[MOË 03] MOËS N., CLOIREC M., CARTRAUD P. *et al.*, "A computational approach to handle complex microstructure geometries", *Computer Methods in Applied Mechanics and Engineering*, vol. 192, pp. 3163–3177, 2003.

[MOO 02] MOORE A., "The circle tree – a hierarchical structure for efficient storage, access and multi-scale representation of spatial data", *SIRC 2002*, Dunedin, New Zealand, pp. 1–8, 2002.

[NEV 95] NEVES M.M., RODRIGUES H., GUEDES J.M., "Generalized topology design of structures with a buckling load criterion", *Structural and Multidisciplinary Optimization*, vol. 10, pp. 71–78, 1995.

[NEV 02] NEVES M.M., SIGMUND O., BENDSØE M.P., "Topology optimization of periodic microstructures with a penalization of highly localized buckling modes", *International Journal for Numerical Methods in Engineering*, vol. 54, pp. 809–834, 2002.

[NIU 88] NIU M.C.Y., *Airframe Structural Design: Practical Design Information and Data on Aircraft Structures*, Technical Book Co, 1988.

[NOC 99] NOCEDAL J., WRIGHT S., *Numerical Optimization*, Springer, 1999.

[NOR 09] NOROUZI Y., RAHMATI S., HOJJAT Y., "A novel lattice structure for SL investment casting patterns", *Rapid Prototyping Journal*, vol. 15, no. 4, pp. 255–263, 2009.

[OH 97] OH J.H., KIM Y.G., LEE D.G., "Optimum bolted joints for hybrid composite materials", *Composite Structures*, vol. 38, no. 1, pp. 329–341, 1997.

[OIN 10] OINONEN A., TANSKANEN P., BJÖRK T. *et al.*, "Pattern optimization of eccentrically loaded multi-fastener joints", *Structural and Multidisciplinary Optimization*, vol. 40, 1–6, pp. 597–609, 2010.

[OLH 05] OLHOFF N., DU J.B., "Topological design of continuum structures subjected to forced vibration", *Proceedings of 6th World Congresses of Structural and Multidisciplinary Optimization*, Rio de Janeiro, Brazil, pp. 1–8, 2005.

[OSH 88] OSHER S., SETHIAN J.A., "Fronts propagating with curvature-dependent speed: algorithms based on Hamilton–Jacobi formulations", *Journal of Computational Physics*, vol. 79, pp. 12–49, 1988.

[PAG 12] PAGNACCO E., LAMBERT S., KHALIJ L. *et al.*, "Design optimisation of linear structures subjected to dynamic random loads with respect to fatigue life", *International Journal of Fatigue*, vol. 43, pp. 168–177, 2012.

[PÁL 06] PÁL L., "A genetic algorithm for the two-dimensional single large object placement problem", *Proceedings of 3rd Romanian–Hungarian Joint Symposium on Applied Computational Intelligence*, pp. 1–11, 2006.

[PEC 99] PECULLAN S., GIBIANSKY L., TORQUATO S., "Scale effects on the elastic behavior of periodic and hierarchical two-dimensional composites", *Journal of the Mechanics and Physics of Solids*, vol. 47, pp. 1509–1542, 1999.

[PED 98] PEDERSEN P., TORTORELLI D., "Constitutive parameters and their evolution", *Control and Cybernetics*, vol. 27, pp. 295–310, 1998.

[PED 00] PEDERSEN N.L., "Maximization of eigenvalues using topology optimization", *Structural and Multidisciplinary Optimization*, vol. 20, pp. 2–11, 2000.

[PED 12] PEDERSEN P., PEDERSEN N.L., "Interpolation/penalization applied for strength design of 3D thermoelastic structures", *Structural and Multidisciplinary Optimization*, vol. 45, pp. 773–786, 2012.

[POO 95] POON C., XIONG Y., *Design of Bolted Joints for Composite Structures*, Woodhead Publishing Limited, UK, pp. 629–636, 1995.

[POO 07] POON N.M., MARTINS J.R., "An adaptive approach to constraint aggregation using adjoint sensitivity analysis", *Structural and Multidisciplinary Optimization*, vol. 34, pp. 61–73, 2007.

[QIA 04] QIAN Z.Y., ANANTHASURESH G.K., "Optimal embedding of rigid objects in the topology design of structures", *Mechanics Based Design of Structures and Machines*, vol. 32, pp. 165–193, 2004.

[QIA 09] QIAO H.T., LIU S.T., "Concurrent optimum design of components layout and connection in a structure", *Chinese Journal of Theoretical and Applied Mechanics*, vol. 41, no. 2, pp. 222–228, 2009.

[QUE 00] QUERIN O.M., YOUNG V., "Computational efficiency and validation of bi-directional evolutionary structure optimization", *Computer Methods in Applied Mechanics and Engineering*, vol. 189, pp. 559–573, 2000.

[QUI 94] QUINLAN S., "Efficient distance computation between non-convex objects", *Proceedings of International Conference on Robotics and Automation*, pp. 3324–3329, 1994.

[RAD 02] RADOVCIC Y., REMOUCHAMPS A., "BOSS Quattro: an open system for parametric design", *Structural and Multidisciplinary Optimization*, vol. 23, pp. 140–152, 2002.

[RAM 11] RAMANI A., "Multi-material topology optimization with strength constraints", *Structural and Multidisciplinary Optimization*, vol. 43, pp. 597–615, 2011.

[REI 07] REICH G.W., SANDERS B., JOO J.J., "Development of skins for morphing aircraft applications via topology optimization", *Journal of Intelligent Material Systems and Structures*, vol. 20, pp. 1–13, 2007.

[REM 11] REMOUCHAMPS A., BRUYNEEL M., FLEURY C. *et al.*, "Application of a bi-level scheme including topology optimization to the design of an aircraft pylon", *Structural and Multidisciplinary Optimization*, vol. 44, pp. 739–750, 2011.

[ROD 95] RODRIGUES H., FERNANDES P., "A material based model for topology optimization of thermoelastic structure", *International Journal for Numerical Methods in Engineering*, vol. 38, pp. 1951–1965, 1995.

[ROD 02] RODRIGUES H., GUEDES J.M., BENDSOE M.P., "Hierarchical optimization of material and structure", *Structural and Multidisciplinary Optimization*, vol. 24, pp. 1–10, 2002.

[RON 00] RONG J.H., XIE Y.M., YANG X.Y. *et al.*, "Topology optimization of structures under dynamic response constraints", *Journal of Sound and Vibration*, vol. 234,pp. 177–189, 2000.

[RON 01] RONG J.H., XIE Y.M., "An improved method for evolutionary structure optimization against buckling", *Computer and Structure*, vol. 79, pp. 253–263, 2001.

[RON 08] RONG J.H., LIANG Q.Q., "A level set method for topology optimization of continuum structures with bounded design domains", *Computer Methods in Applied Mechanics and Engineering*, vol. 197, pp. 1447–1465, 2008.

[ROW 82] ROWLANDS R., RAHMAN M., WILKINSON T. *et al.*, "Single-and multiple-bolted joints in orthotropic materials", *Composites*, vol. 13, no. 3, pp. 273–279, 1982.

[ROZ 01a] ROZVANY G.I.N., "Aims, scope, methods, history and unified terminology of computer-aided topology optimization in structural mechanics", *Structural and Multidisciplinary Optimization*, vol. 21, pp. 90–108, 2001.

[ROZ 01b] ROZVANY G.I.N., "Stress ratio and compliance based methods in topology optimization – a critical review", *Structural and Multidisciplinary Optimization*, vol. 21, pp. 109–119, 2001.

[ROZ 02a] ROZVANY G.I.N., QUERIN O.M., "Combining ESO with rigorous optimality criteria", *International Journal of Vehicle Design*, vol. 28, pp. 294–299, 2002.

[ROZ 02b] ROZVANY G.I.N., QUERIN O.M., "Theoretical foundations of sequential element rejections and admissions (SERA) methods and their computational implementations in topology optimization", *Proceedings of the 9th AIAA/ISSMO Symposium on Multidisciplinary Analysis and Optimization*, AIAA, Reston, VA, pp. 1–4, 2002.

[ROZ 04] ROZVANY G.I.N., QUERIN O.M., "Sequential element rejections and admissions (SERA) method: applications to multiconstraint problems", *Proceedings of the 10th AIAA/ISSMO Multidisciplinary Analysis Optimization Conference*, Albany, NY, 2004.

[SAM 89] SAMET H., *Spatial Data Structures: Quadtree, Octree and Other Hierarchical Methods*, Addison Wesley, 1989.

[SHI 11] SHI Y., JIANG Y., WANG Y., "Application and improvement of direct solving method in seismic response analysis of structures under multi-support excitations", *Engineering Mechanics*, vol. 28, pp. 75–81, 2011.

[SHU 11] SHU L., WANG M.Y., FANG Z. *et al.*, "Level set based structural topology optimization for minimizing frequency response", *Journal of Sound and Vibration*, vol. 330, no. 24, pp. 5820–5834, 2011.

[SIG 94] SIGMUND O., "Materials with prescribed constitutive parameters: an inverse homogenization problem", *International Journal of Solids and Structures*, vol. 31, pp. 2313–2329, 1994.

[SIG 95] SIGMUND O., "Tailoring materials with prescribed elastic properties", *Mechanics of Materials*, vol. 20, pp. 351–368, 1995.

[SIG 97] SIGMUND O., TORQUATO S., "Design of materials with extreme thermal expansion using a three-phase topology optimization method", *Journal of the Mechanics and Physics of Solids*, vol. 45, pp. 1037–1067, 1997.

[SIG 98] SIGMUND O., PETERSSON J., "Numerical instabilities in topology optimization: a survey on procedures dealing with checkerboards, mesh-dependencies and local minima", *Structural and Multidisciplinary Optimization*, vol. 16, pp. 68–75, 1998.

[SIG 99] SIGMUND O., TORQUATO S., "Design of smart composite materials using topology optimization", *Smart Materials and Structures*, vol. 8, pp. 365–379, 1999.

[SIG 01a] SIGMUND O., "A 99 line topology optimization code written in MATLAB", *Structural and Multidisciplinary Optimization*, vol. 21, pp. 120–127, 2001.

[SIG 01b] SIGMUND O., "Design of multiphysics actuators using topology optimization – Part II: two-material structures", *Computer Methods in Applied Mechanics Engineering*, vol. 190, pp. 6605–6627, 2001.

[SIG 06] SIGMUND O., "On topology optimization with manufacturing constraints", *3rd European Conference on Computational Mechanics Solids, Structures and Coupled Problems in Engineering*, Lisbon, Portugal, 2006.

[SIG 07] SIGMUND O., "Morphology-based black and white filters for topology optimization", *Structural and Multidisciplinary Optimization*, vol. 33, pp. 401–424, 2007.

[SIG 13] SIGMUND O., MAUTE K., "Topology optimization approaches", *Structural and Multidisciplinary Optimization*, vol. 48, pp. 1031–1055, 2013.

[SQU 98] SQUIRE W., TRAPP G., "Using complex variables to estimate derivatives of real functions", *SIAM Review*, vol. 40, pp. 110–112, 1998.

[STE 05] STEGMANN J., LUND E., "Discrete material optimization of general composite shell structures", *International Journal for Numerical Methods in Engineering*, vol. 62, pp. 2009–2027, 2005.

[STO 01] STOLPE M., SVANBERG K., "An alternative interpolation scheme for minimum compliance topology optimization", *Structural and Multidisciplinary Optimization*, vol. 22, pp. 116–124, 2001.

[SUT 99] SUTHERLAND L., SHENOI R., LEWIS S., "Size and scale effects in composites: I. Literature review", *Composites Science and Technology*, vol. 59, pp. 209–220, 1999.

[SUZ 91] SUZUKI K., KIKUCHI N., "A homogenization method for shape and topology optimization", *Computer Methods in Applied Mechanics and Engineering*, vol. 93, pp. 291–318, 1991.

[SVA 87] SVANBERG K., "The method of moving asymptotes – a new method for structural optimization", *International Journal for Numerical Methods in Engineering*, vol. 24, pp. 359–373, 1987.

[SVA 95] SVANBERG K., "A globally convergent version of MMA without line search", *1st World Congress of Structural and Multidisciplinary Optimization*, Pergamon, New York, 1995.

[SVA 07] SVANBERG K., "On a globally convergent version of MMA", *7th World Congress on Structural and Multidisciplinary Optimization*, Seoul, 2007.

[TAN 02] TANSKANEN P., "The evolutionary structural optimization method: theoretical aspects", *Computer Methods in Applied Mechanics and Engineering*, vol. 191, pp. 5485–5498, 2002.

[TAN 05] TANTIKOM K., AIZAWA T., MUKAI T., "Symmetric and asymmetric deformation transition in the regularly cell-structured materials. Part I: experimental study", *International Journal of Solids and Structures*, vol. 42, pp. 2199–2210, 2005.

[TCH 02] TCHERNIAK D., "Topology optimization of resonating structures using SIMP method", *International Journal for Numerical Methods in Engineering*, vol. 54, pp. 1605–1622, 2002.

[THO 92] THOMSEN J., "Topology optimization of structures composed of one or two materials", *Structural and Multidisciplinary Optimization*, vol. 5, pp. 108–115, 1992.

[TOM 11] TOMLIN M., MEYER J., "Topology optimization of an additive layer manufactured ALM aerospace part", *The 7th Altair CAE Technology Conference*, 2011.

[VAN 99] VAN HUMBEECK J., "Non-medical applications of shape memory alloys", *Materials Science and Engineering A – Structural*, vol. 273, pp. 134–148, 1999.

[VAN 07] VAN MIEGROET L., DUYSINX P., "Stress concentration minimization of 2D filets using X-FEM and level set description", *Structural and Multidisciplinary Optimization*, vol. 33, pp. 425–438, 2007.

[WAN 88] WANG S.M., HAN Y.X., "Finite element analysis for load distribution of multi-fastener joints", *Journal of Composite Materials*, vol. 22, no. 2, pp. 124–135, 1988.

[WAN 00] WANG B.Z., *Aircraft Design Manual 10: Structural Design*, Aviation Industry Press, Beijing, 2000.

[WAN 03] WANG M.Y., WANG X., GUO D., "A level set method for structural topology optimization", *Computer Methods in Applied Mechanics and Engineering*, vol. 192, pp. 227–246, 2003.

[WAN 05] WANG M.Y., WANG S.Y., "Bilateral filtering for structural topology optimization", *International Journal for Numerical Methods in Engineering*, vol. 63, pp. 1911–1938, 2005.

[WEI 10] WEI P., WANG M.Y., XING X., "A study on X-FEM in continuum structural optimization using a level set model", *Computer-Aided Design*, vol. 42, pp. 708–719, 2010.

[WIJ 09] WIJKER J.J., Random Vibrations in Spacecraft Structures Design: Theory and Applications, Springer, 2009.

[WIL 82] WILSON E.L., YUAN M.W., DICKENS J.M., "Dynamic analysis by direct superposition of Ritz vectors", *Earthquake Engineering and Structural Dynamics*, vol. 10, pp. 813–821, 1982.

[XIE 94a] XIE Y.M., STEVEN G.P., "Optimal design of multiple load case structures using an evolutionary procedure", *Engineering Computations*, vol. 11, pp. 295–302, 1994.

[XIE 94b] XIE Y.M., STEVEN G.P., "A simple approach to structure frequency optimization", *Computer and Structure*, vol. 53, pp. 1487–1491, 1994.

[XIE 96] XIE Y.M., STEVEN G.P., "Evolutionary structure optimization for dynamic problems", *Computer and Structure*, vol. 58, pp. 1067–1073, 1996.

[XIE 07] XIE K., WELLS L., CAMELIO J.A. *et al.*, "Variation propagation analysis on compliant assemblies considering contact interaction", *Journal of Manufacturing Science and Engineering – Transactions of the ASME*, vol. 129, pp. 934–942, 2007.

[YAN 99a] YANG X.Y., XIE Y.M., STEVEN G.P., "Bidirectional evolutionary method for stiffness optimization", *AIAA Journal*, vol. 37, pp. 1483–1488, 1999.

[YAN 99b] YANG X.Y., XIE Y.M., STEVEN G.P., "Topology optimization for frequency using an evolutionary method", *Journal of Structural Engineering*, vol. 125, pp. 1432–1438, 1999.

[YAN 02] YANG X.Y., XIE Y.M., LIU J.S. *et al.*, "Perimeter control in the bidirectional evolutionary optimization method", *Structural and Multidisciplinary Optimization*, vol. 24, pp. 430–440, 2002.

[YAN 08] YAN J., CHENG G.D., LIU L., "A uniform optimum material based model for concurrent optimization of thermoelastic structures and materials", *International Journal for Simulation and Multidisciplinary Design Optimization*, vol. 2, pp. 259–266, 2008.

[YAN 14] YANG X., LI Y., "Structural topology optimization on dynamic compliance at resonance frequency in thermal environments", *Structural and Multidisciplinary Optimization*, vol. 49, pp. 81–91, 2014.

[YIN 01] YIN L., ANANTHASURESH G.K., "Topology optimization of compliant mechanisms with multiple materials using a peak function material interpolation scheme", *Structural and Multidisciplinary Optimization*, vol. 23, pp. 49–62, 2001.

[YOO 04] YOON K.H., HEO S.P., SONG K.N. *et al.*, "Dynamic impact analysis of the grid structure using multi-point constraint (MPC) equation under the lateral impact load", *Computers and Structures*, vol. 82, pp. 2221–2228, 2004.

[YOO 10] YOON G.H., "Structural topology optimization for frequency response problem using model reduction schemes", *Computer Methods in Applied Mechanics and Engineering*, vol. 199, no. 25, pp. 1744–1763, 2010.

[YOO 11] YOON G.H., "Topology optimization for nonlinear dynamic problem with multiple materials and material-dependent boundary condition", *Finite Elements in Analysis and Design*, vol. 47, pp. 753–763, 2011.

[ZHA 96] ZHAO M.Y., YANG L., WAN X.P., "Load distribution in composite multifastener joints", *Computers and Structures*, vol. 60, no. 2, pp. 337–342, 1996.

[ZHA 97] ZHANG W.H., FLEURY C., "A modification of convex approximation methods for structural optimization", *Computers and Structures*, vol. 64, pp. 89–95, 1997.

[ZHA 03] ZHANG W.H., DUYSINX P., "Dual approach using a variant perimeter constraint and efficient sub-iteration scheme for topology optimization", *Computer and Structure*, vol. 81, pp. 2173–2181, 2003.

[ZHA 06] ZHANG W.H., SUN S.P., "Scale-related topology optimization of cellular materials and structures", *International Journal for Numerical Methods in Engineering*, vol. 68, pp. 993–1011, 2006.

[ZHA 07] ZHANG W.H., DAI G.M., WANG F.W. *et al.*, "Using strain energy-based prediction of effective elastic properties in topology optimization of material microstructures", *Acta Mechanica Sinica*, vol. 23, pp. 77–89, 2007.

[ZHA 10a] ZHANG Q., ZHANG W.H., ZHU J.H., "Topology optimization of structures under dynamic response constraints", *Journal of Mechanical Engineering*, vol. 46, pp. 45–51, 2010.

[ZHA 10b] ZHANG Y.W., LIN J.H., ZHAO Y. *et al.*, "Symplectic random vibration analysis of a vehicle moving on an infinitely long periodic track", *Journal of Sound and Vibration*, vol. 329, no. 21, pp. 4440–4454, 2010.

[ZHA 11] ZHANG W.H., XIA L., ZHU J.H., "Some recent advances in the integrated layout design of multicomponent systems", *Journal of Mechanical Design*, vol. 133, no. 10, pp 104503, 2011.

[ZHA 12] ZHANG Q., ZHANG W.H., ZHU J.H. *et al.*, "Layout optimization of multi-component structures under static loads and random excitations", *Engineering Structures*, vol. 43, pp. 120–128, 2012.

[ZHA 14a] ZHANG W., YANG J., XU Y. *et al.*, "Topology optimization of thermoelastic structures: mean compliance minimization or elastic strain energy minimization", *Structural and Multidisciplinary Optimization*, vol. 49, pp. 417–429, 2014.

[ZHA 14b] ZHANG W.H., ZHANG Z.D., ZHU J.H. *et al.*, "Structural topology optimization: extensibility and attainability", *Science China – Technological Sciences*, vol. 57, pp. 1310–1321, 2014.

[ZHA 15] ZHANG W.H., LIU H., GAO T., "Topology optimization of large-scale structures subjected to stationary random excitation: an efficient optimization procedure integrating pseudo excitation method and mode acceleration method", *Computers and Structures*, vol. 158, pp. 61–70, 2015.

[ZHO 91] ZHOU M., ROZVANY G.I.N., "The COC algorithm, part II: topological, geometry and generalized shape optimization", *Computer Methods in Applied Mechanics and Engineering*, vol. 89, pp. 197–224, 1991.

[ZHO 01] ZHOU M., ROZVANY G.I.N., "On the validity of ESO type methods in topology optimization", *Structural and Multidisciplinary Optimization*, vol. 21, pp. 80–83, 2001.

[ZHO 04] ZHOU M., "Topology optimization for shell structures with linear buckling responses", *World Congress of Computational Mechanics*, Beijing, China, 2004.

[ZHO 16] ZHOU Y., ZHANG W.H., ZHU J.H. *et al.*, "Feature-driven topology optimization method with signed distance function", *Computer Methods in Applied Mechanics and Engineering*, vol. 310, pp. 1–32, 2016.

[ZHU 06a] ZHU J.H., ZHANG W.H., "Maximization of structure natural frequency with optimal support layout", *Structural and Multidisciplinary Optimization*, vol. 31, pp. 462–469, 2006.

[ZHU 06b] ZHU J.H., ZHANG W.H., "Coupled design of components layout and supporting structures using shape and topology optimization", *Proceedings of the Fourth China-Japan-Korea Joint Symposium on Optimization of Structural and Mechanical Systems*, Kunming, China, 2006.

[ZHU 07] ZHU J.H., ZHANG W.H., QIU K.P., "Bi directional evolutionary topology optimization using element replaceable method", *Computational Mechanics*, vol. 40, pp. 97–109, 2007.

[ZHU 09] ZHU J.H., ZHANG W.H., BECKERS P., "Integrated layout design of multi-component system", *International Journal for Numerical Methods in Engineering*, vol. 78, pp. 631–651, 2009.

[ZHU 10a] ZHU J.H., ZHANG W.H., "Integrated layout design of supports and structures", *Computer Methods in Applied Mechanics and Engineering*, vol. 199, pp. 9–12, 557–569, 2010.

[ZHU 10b] ZHUANG C., XIONG Z., DING H., "Topology optimization of multi-material for the heat conduction problem based on the level set method", *Engineering Optimization*, vol. 42, pp. 811–831, 2010.

[ZHU 16] ZHU J.H., ZHANG W.H., XIA L., "Topology optimization in aircraft and aerospace structures design", *Archives of Computational Methods in Engineering*, 2016.

# Index

## A, B, C, D



## E, F, H



## I, L, M

## P, R, S



## T, X

FSC
www.fsc.org
MIX
Papier | Fördert gute Waldnutzung
FSC® C083411

Kolibri 360 GmbH
Ferdinand-Jühlke-Straße 7
99095 Erfurt, Deutschland
produktsicherheit@kolibri360.de